INDUSTRIAL MAINTENANCE AND TROUBLESHOOTING

FOURTH EDITION

WORKBOOK

AMERICAN TECHNICAL PUBLISHERS
Orland Park, Illinois

The publisher is grateful for the photographs provided by the following companies and organizations.

DPSI (DP Solutions, Inc.)
Integrus Architecture
Miller Fall Protection
North Safety Products

4 5 6 7 8 9 – 19 – 9 8 7 6 5 4 3 2

Printed in the United States of America

ISBN 978-0-8269-3687-5

This book is printed on recycled paper.

Table of Contents

5 Mechanical Drives

6 Electrical Systems

7 Electronic Systems

8 Refrigeration Systems

9 Boiler Systems

10 HVAC Systems

11 Fluid Power Systems

12 Efficiency and Sustainability

MAINTENANCE PERSONNEL

SECTION **1.1**

REVIEW QUESTIONS

Name ______________________ Date ______________

True-False

T F 1. Commercial buildings include facilities for food processing, manufacturing, refining, mining, assembly, and other heavy processes.

T F 2. Multiskilled maintenance personnel have a great amount of expertise in one craft.

T F 3. Various maintenance personnel jobs are defined by the federal Standard Occupational Classification (SOC) system and identified by job titles and numbers.

T F 4. When dealing with an angry customer, explain the possible solutions and help the customer understand the pros and cons of each possible solution.

T F 5. Successful maintenance personnel practice lifelong learning and continuous improvement.

T F 6. Maintenance personnel make the most of each task by evaluating and refining strategies and skills.

T F 7. Networking can only be local (inside the building), with equipment being monitored within the location.

T F 8. Clear and concise communication is the key to effective customer service during maintenance and troubleshooting.

Multiple Choice

______________ 1. In an industrial facility, downtime due to equipment maintenance or breakdowns can be ___.

A. used for team-building exercises
B. irrelevant
C. a way to save money
D. extremely costly

______________ 2. When dealing with an angry customer, ___.

A. never allow them to vent their frustrations
B. allow them to vent their frustrations
C. only speak when spoken to
D. file a complaint with your supervisor

______________ **3.** Once a repair is completed, technicians must ___.
- A. carefully record the problem, causes, and solutions
- B. diagnose the cause
- C. troubleshoot the problem
- D. confirm the customer's understanding of the problem

Completion

______________ **1.** ___ maintenance personnel have expertise in several crafts.

______________ **2.** ___ skills are strategies and actions that allow an individual to communicate effectively with others in a variety of situations.

______________ **3.** Effective ___ requires active listening followed by careful speaking.

UNSCHEDULED MAINTENANCE

SECTION **1.2**

REVIEW QUESTIONS

Name ______________________________ **Date** ____________________

True-False

T F **1.** Preventive maintenance includes emergency work and breakdown maintenance, which differ based on whether the failed equipment receives regular maintenance.

T F **2.** Emergency work orders are issued to repair damaged equipment immediately.

T F **3.** Breakdown maintenance is the most sophisticated maintenance work and is often appropriate for equipment that is inexpensive and noncritical to facility operations.

Multiple Choice

______________ **1.** ___ work is unscheduled service to correct an unexpected failure on equipment that receives a regular maintenance.

A. Preventive
B. Breakdown
C. Emergency
D. Periodic

______________ **2.** Light bulbs are serviced using ___ maintenance because it is less costly to replace a bad bulb than to predict a bulb failure.

A. emergency
B. periodic
C. preventive
D. breakdown

Completion

______________ **1.** ___ maintenance is immediate service that is required due to a failure.

______________ **2.** Keeping a(n) ___ of emergency work provides information that can improve maintenance procedures or equipment design by identifying common equipment problems.

______________ **3.** ___ maintenance is unscheduled service on failed equipment that has not received regular maintenance.

PREVENTIVE MAINTENANCE

SECTION **1.3**

REVIEW QUESTIONS

Name ______________________________ **Date** ____________________

True-False

(T)	F	**1.** PM tasks and frequency for each piece of equipment are determined using manufacturer specifications, equipment manuals, trade publications, and worker experience. P10
T	(F)	**2.** Breakdown maintenance is work completed at specific intervals. P 10
(T)	F	**3.** Project work commonly includes rebuilding or modifying equipment, renovating structures, or installing new equipment. P 11
(T)	F	**4.** Maintenance technicians are an important part of the changeover of equipment to different product lines because most problems occur at this time.
T	(F)	**5.** A startup process is a system used to record and organize maintenance information, which is then used to make the decisions required to maintain the facility and equipment. P 10
(T)	F	**6.** Data from a plant survey can be entered into a PM system to create a master file for each piece of equipment. P 13
(T)	F	**7.** Electronic files are becoming more popular for storing maintenance information because they are easy to store, organize, and access. P14
T	(F)	**8.** The first place to look for specific PM task recommendations is in the logbook for each piece of equipment. P 15

Multiple Choice

___A___ **1.** ___ maintenance is scheduled work required to keep equipment in peak operating condition.

A. Preventive
B. Calendar
C. Plant
D. Breakdown

P10

D ______ **2.** ___ maintenance is performed to minimize emergency work and ensure reliable and efficient operations. P10

A. Breakdown
B. Priority
C. Unscheduled
D. Scheduled

B ______ **3.** With a consistent flow of accurate operation information, ___ can help increase efficiency, reduce costs, and minimize health and safety problems. P13

A. networks
B. PM systems
C. inventory control
D. operational diagrams

Completion

Scheduled ______ **1.** ___ maintenance is work that is planned and scheduled for completion. P10

Collective ______ **2.** ___ work is repair of a known problem before a breakdown occurs. P11

operational ______ **3.** ___ maintenance is the monitoring of running equipment to ensure that it is functioning correctly. P11

alarms ______ **4.** Automated control systems can activate ___, such as sounds, indicator lights, or error codes, if problems arise. P13

Survey ______ **5.** A plant ___ is a complete inventory and condition assessment of a facility's equipment and structure. P13

PREDICTIVE MAINTENANCE

SECTION 1.4

REVIEW QUESTIONS

Name ______________________________ **Date** ____________________

True-False

p17 Predictive

T F **1.** Preventive maintenance is a system of looking for symptoms of a problem that has not caused a failure yet but probably will in the future.

T F **2.** If a measurement exceeds the acceptable tolerance, even if the equipment continues to operate adequately, then corrective work is scheduled to address the problem. P17

T F **3.** Since a PdM program does not typically require substantial investment, it is commonly used for all equipment. P18

T F **4.** PdM analysis may indicate that failure is approaching, but it is likely impossible to know the exact timing of the potential breakdown. P19

T F **5.** A PdM program can be overly conservative, and useful service life of the equipment is wasted when it is repaired or replaced too early. P19

T F **6.** The simplest type of PdM uses manufacturer statistics on the average life span of the equipment. P17

T F **7.** Unusual appearance or sounds of operating equipment may be obvious to trained maintenance personnel and are signs of a problem that is hard to detect by other means. P18

T F **8.** Scheduled monitoring may be done whenever all time-sensitive work is completed and technicians are available for PdM work. P18

Multiple Choice

D **1.** ___ maintenance is the monitoring of equipment operation characteristics to determine the degree of wear in a system, compare them to normal tolerances, and predict potential malfunctions or failures.

A. Periodic
B. Corrective
C. Preventive
D. Predictive

P17

______ **2.** In addition to temperature analysis, other common types of PdM programs include ___ analysis.

A. vibration and ultrasonic
B. gas and vapor
C. leak
D. radiation and chemical

______ **3.** ___ monitoring requires sensing devices that are permanently attached to the equipment being monitored.

A. Random
B. Scheduled
C. Continuous
D. Unscheduled

Completion

______ **1.** ___ monitoring is unscheduled equipment monitoring as required.

______ **2.** ___ monitoring is equipment monitoring at specific time intervals.

______ **3.** ___ monitoring is equipment monitoring at all times.

MAINTENANCE AND REPAIR MANAGEMENT

SECTION **1.5**

REVIEW QUESTIONS

Name ______________________ **Date** ______________

True-False

T	F	**1.** Work orders are used to organize, plan, and monitor scheduled maintenance tasks.
T	F	**2.** All work orders have a flexible completion date.
T	F	**3.** Work priority is indicated on a work order.
T	F	**4.** The highest priority is for scheduled periodic maintenance.
T	F	**5.** CMMSs are used to collect a variety of information from many sources, organize and store this information, and provide data analysis as needed.
T	F	**6.** The most common mobile device used for service and troubleshooting in the field is a walkie-talkie.
T	F	**7.** A photo of equipment is extremely useful for reducing downtime when replacing components.
T	F	**8.** The cost of ordering parts includes both the cost of the parts and the time required for placing the order and sorting the parts when they arrive.
T	F	**9.** The most direct types of failure costs are the costs to repair or replace failed equipment.
T	F	**10.** When equipment fails, the maintenance department may need to choose between repair or replacement.

Multiple Choice

______________ **1.** A ___ is a document that details specific maintenance tasks to be completed.
- A. logbook
- B. written order
- C. work order
- D. data log

A 2. ___ can be extremely useful in troubleshooting.
A. Logbooks
B. Inventory lists
C. Plant surveys
D. Work priorities

P21

Completion

P 19

Work priority 1. ___ is the order in which work should be done based on importance.

logbook 2. A(n) ___ is a notebook or electronic file that documents maintenance tasks and additional information. P21

CMMS 3. A(n) ___ is a software package or web-based system that organizes PM information and automatically generates reports, work orders, and other data for implementing and improving future maintenance activities. P22

numbered 4. All spare parts should be ___ so their location and quantity can be tracked. P27

repair vs replace 5. The two common ways in which costs are analyzed to determine the best value for a facility are maintenance versus failure costs and ___ costs.

P27

TROUBLESHOOTING

SECTION **1.6**

REVIEW QUESTIONS

Name ____________________ Date ____________________

True-False

T F 1. Troubleshooting methodically eliminates various system processes and components as causes of a malfunction until the malfunction is located and diagnosed.

T F 2. When arriving on-site to investigate a problem, the first step is comparing possible solutions.

T F 3. It is important to look for potential secondary causes, which are problems that cause the conditions that caused the original problem.

T F 4. The wear-out period is the period after the break-in period when most equipment operates as designed.

T F 5. If a repair does not fix the original problem, then the troubleshooting process resumes by reevaluating the written lists of symptoms, causes, and troubleshooting test charts.

T F 6. New system problems can arise when correcting an old problem.

T F 7. While the malfunction of temperature-sensitive equipment could be easily traced to a higher temperature, electronics failures would appear to be random and have no obvious cause.

T F 8. Troubleshooting activities do not generate any information that is useful to a PM system.

Multiple Choice

______ 1. In some cases, equipment displays include ___ or other error indicators.
- A. work orders
- B. fault codes
- C. operator's manuals
- D. flow charts

______ 2. ___ use symbols and interconnecting lines to help a troubleshooter follow a logical path when solving a problem.
- A. Logbooks
- B. Work orders
- C. Testing charts
- D. Flow charts

___A___ **3.** The ___ is the time just after installation when equipment achieves peak operating performance.
A. break-in period
B. useful life
C. wear-out period
D. start-up sequence

___D___ **4.** Whenever possible, equipment is inspected when it is ___.
A. not locked out
B. energized
C. being replaced
D. de-energized

___B___ **5.** ___ is the consideration of an entire system and its interrelationships with other systems while troubleshooting.
A. Computer literacy
B. Systems thinking
C. Preventive maintenance
D. Computerized maintenance management

___D___ **6.** ___ systems become more complicated as the number of input measurements and the sensitivity of the controls increase.
A. Cascade
B. Computerized maintenance management
C. Open-loop
D. Closed-loop

Completion

___Troubleshoting___ **1.** ___ is the systematic investigation of the causes of system problems to determine the best solution.

___testing Chart___ **2.** A(n) ___ is a record of the tests performed, along with their results, during a troubleshooting process.

___flow chart___ **3.** A(n) ___ is a diagram that shows a logical sequence of troubleshooting steps for a given set of conditions.

___Wear-ont___ **4.** The ___ period is the period after the useful life of equipment when normal failures occur.

___troubleshooting report___ **5.** A(n) ___ is a document that includes the symptoms, causes, tests conducted, repairs chosen, and condition of equipment after a repair.

___open-loop___ **6.** A(n) ___ system is a system in which decisions are made based on only the current state of the system and a model of how it should work.

MAINTENANCE AND TROUBLESHOOTING PRINCIPLES

CHAPTER 1

ACTIVITIES

Name ______________________ **Date** ______________

Equipment Life Expectancy Curve

___A___ **1.** Break-in period

___B___ **2.** Useful life

___C___ **3.** Wear-out period

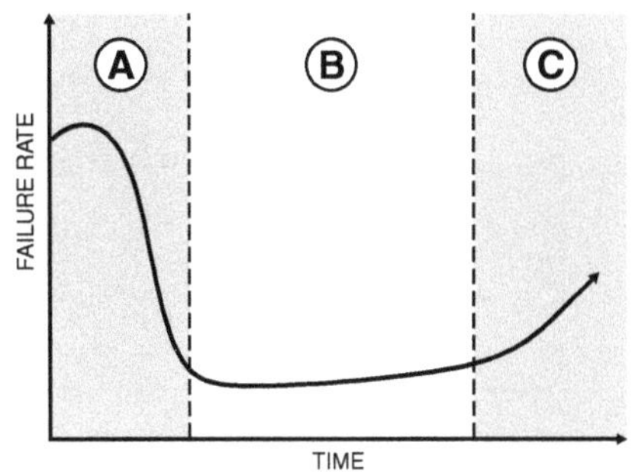

Periodic Maintenance

__________ **1.** Maintenance frequency

__________ **2.** Maintenance task

__________ **3.** Equipment description

__________ **4.** Work order number

__________ **5.** Maintenance procedure

__________ **6.** Previous maintenance work completed

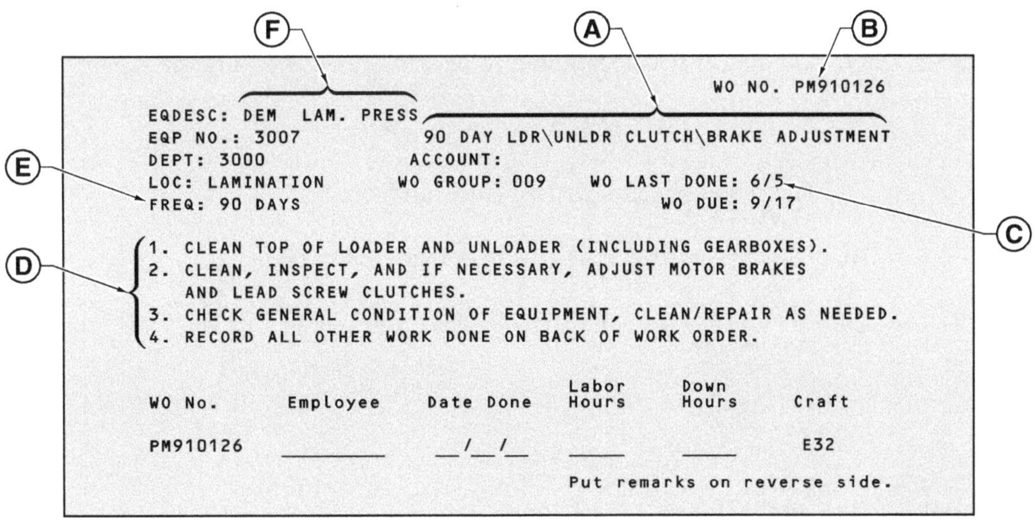

Work Order—Corrective

```
                    WORK ORDER-CORRECTIVE
Work Order:   COR 345               Description:   Repair damaged chain guard
Asset/Equip:   Conveyor #2
Procedure:   Repair damaged chain guard
Requested by:   Packing Lead
Telephone:   Ext. 556
Asset Shutdown:   X   Plant Shutdown
Model:   NA
Serial No:   NA
Location:   Packing Room
Skill:   Level 2
Sch Date:   Timely
Priority:   2
Shift:   NA
Status:   Open
Labor     Craft     Labor Description     Est Hrs. REF OVE DOUBL
    Maintenance     (Name printed in)
Tasks:   ID COR 345 PR 2
   Notify Packing Lead Person about repair.
   Lock out power to packing line #2.
   Repair damaged chain guard or fabricate new guard on
        drive motor for conveyor #2 in packing area.
   Remove Lockout.
   Notify Packing Lead Person when completed.
Comments: 2 hours.
          Damage caused by forklift.
          Rebuilt guard from mounts out.
          No unusual costs.
          Need to order more 6010 rod.
```

__________ **1.** A(n) ___ is damaged on Conveyor #2.

__________ **2.** The repair to Conveyor #2 required ___ hours.

__________ **3.** Conveyor #2 is located in the ___ Room.

__________ **4.** The ___ is to be notified when the repair is completed.

A. superintendent
B. supervisor
C. packing lead person
D. maintenance manager

__________ **5.** The priority level for this work order is ___.

__________ **6.** The damage was caused by a(n) ___.

T F **7.** The power was locked out to packing line #2 during the repair.

T F **8.** Additional 6012 rods are to be ordered.

T F **9.** The repair necessitated a plant shutdown.

T F **10.** The costs for the repair were considered to be usual.

Master Equipment File

Description:	Furnace #1	Voltage:	440	Warranty ID:	IT6P882
Asset ID:	FUR0001.00	Amperage:	600	Warranty Date:	02/17
Asset Type:	Furnace Systems	Wattage:	264000		
Parent ID:		Phase:	3	YTD Labor Hr:	45.00
Priority:	8 Active:☒	Elec Line:	10	YTD Downtime:	22.00
Manufacturer:	Brown Boveri, Inc.	Air Area:	COMP 6		
Model:	IT6P			TD Labor Hr:	740.00
Serial Number:	OP2810C2B			TD Downtime:	493.00
Vendor:	Lewis Systems, Inc.	Counter UOM:			
Vendor Address:	1862 Erie St.	Current Counter:	1	YTD Labor Cost:	724.00
	Cleveland, OH 55117	Counter Rollover:	0	YTD Misc Cost:	1231.22
Vendor Phone:	216-555-1340	Meter UOM:		YTD Part Cost:	7701.19
Asset Tag:	00509	Current Meter:	1234550	Total:	9656.41
Location:	4247 Piedmont Building 2				
	Floor-1 Room 23 CL15F			TD Labor Cost:	9620.60
Department ID:		Meter Rollover:	0	TD Misc Cost:	5631.80
Cost Center:	Fixed Asset Repair	Purchase Date:	02/17	TD Part Cost:	14,942.00
Supervisor:	Jones, Fred	Install Date:	08/05	Total:	30,194.40
		Retire Date:			
		Install Cost:	97000		
		Replacement Cost:	97000		

Comment: Manufact. warranty extremely strict. Document all hours worked and parts used.

Report Totals:						
	YTD Labor Hr:	45.00	YTD Labor Cost:	724.00	TD Labor Cost:	9620.60
Assets: 1	YTD Downtime:	22.00	YTD Misc Cost:	1231.22	TD Misc Cost:	5631.80
	TD Labor Hr:	740.00	YTD Part Cost:	7701.19	TD Part Cost:	14,942.00
	TD Downtime:	493.00	Total:	9656.41	Total:	30,194.40

DPSI (DP Solutions, Inc.)

MASTER EQUIPMENT FILE

______________ **1.** Furnace #1 is located in Room ___ of Building 2.

______________ **2.** Furnace #1 has a total of ___ hr of labor YTD.

______________ **3.** The installed cost of the furnace was $___.

______________ **4.** The calculated hourly cost for labor YTD is $___.

______________ **5.** The calculated hourly cost for labor TD is $___.

T F **6.** There is a manufacturer's warranty for Furnace #1.

T F **7.** The furnace operates at 220 V.

______________ **8.** The replacement cost is ___.

A. less than the installation cost
B. the same as the installation cost
C. more than the installation cost
D. not given

______________ **9.** The total labor, misc, and part cost TD is $___.

A. 5631.80
B. 9620.60
C. 14,942.00
D. 30,194.40

T F **10.** The electrical service for the furnace is 1ϕ.

Activity: Preventive Maintenance

The manufacturer's maintenance literature recommends cleaning an HVAC system's cooling coils every three months.

1. If the coils are found to be clean every time they are inspected, would it be possible to lengthen the cleaning time to six or nine months?

2. What economic benefit would this have?

A production line includes both an electric heating element and a photoelectric sensor.

3. Which device would likely require more preventive maintenance?

4. Why could the heating element be run on breakdown maintenance?

One production line runs four days a week, eight hours a day. Another production line runs five days a week, 24 hours a day.

5. When establishing a new PM program, which production line is likely the most cost-effective one to begin with?

Periodic maintenance schedules must be established for several pieces of equipment.

6. For a brand-new piece of equipment, what should be the chief source of information when establishing the maintenance schedule?

7. For an older piece of equipment that has little or no manufacturer documentation, what sources of information could be used to establish the maintenance schedule?

8. List several advantages of a CMMS system over a paper-and-file-based PM system.

9. Why must maintenance personnel complete the exact maintenance required for new equipment?

Activity: Maintenance Scenarios

Answer the following questions about maintenance activities.

1. When walking by a hydraulic system, what impromptu inspections can be done?

2. When is the best time to search for leaks in a compressed air supply system?

3. Why is the refrigerator in the staff lunchroom run on breakdown maintenance but the refrigerator that stores chemicals used in oil analysis is inspected and cleaned on a regularly scheduled basis?

4. What should be done if filters are found to be missing during the survey inspection of an air handler?

Activity: Predictive Maintenance

Answer the following questions about predictive maintenance analysis methods.

1. What are the general characteristics of machinery that could benefit from predictive maintenance?

2. Why is wear particle analysis important for old equipment?

3. Why must strict oil sampling procedures be followed?

4. What should be done if a hot spot is found during a thermography inspection of a circuit breaker?

Vibration analysis determines that the bearings on a large, expensive motor are beginning to fail, so work is scheduled for a time when the motor is not needed for production.

5. What type of maintenance is this?

Activity: Troubleshooting Process

Explain the general purpose of each step of the troubleshooting process.

1. Investigating.

2. Isolating.

3. Remedying.

4. Documenting.

5. At which step of the troubleshooting process are interpersonal skills most important?

Activity: Work Orders

An old motor drive was already scheduled to be replaced with new equipment the following month when it suddenly malfunctioned and disrupted production.

1. Discuss the factors that help determine what should be done about the motor drive.

A maintenance technician is paged twice within a short time to respond to two different production machines that have malfunctioned.

2. What factors should be considered when deciding which call to answer first?

During a PM inspection, it is discovered that a pump needs replacement.

3. What type of maintenance work order will be generated?

SAFETY PLANNING

SECTION 2.1

REVIEW QUESTIONS

Name Ethan O Date 2/23/24

True-False

(T) F 1. Working safely is a habit established through training, development of technical skills, and work experience.

(T) F 2. Maintenance personnel must anticipate the necessary safety equipment and procedures to deal with weather conditions, work area conditions, tasks, and facility conditions that may change quickly.

Multiple Choice

B 1. The ___ is responsible for following and enforcing applicable safety regulations, providing necessary safety equipment and training, and reinforcing a workplace culture of safety awareness.

- A. local jurisdiction
- B. employer
- C. employee
- D. technical society

D 2. Safety planning begins with ___.

- A. wearing proper personal protective equipment
- B. lockouts and tagouts
- C. reading the equipment manual
- D. identifying the specific hazards of a machine or process

Completion

Safety 1. ___ is a primary concern during all maintenance work.

Work orders 2. All ___ must contain a section for identifying safety concerns.

Employee 3. The ___ is responsible for following applicable safety regulations, using equipment and following procedures properly, and promptly reporting any safety concerns to supervisors.

SAFETY REGULATIONS

SECTION **2.2**

REVIEW QUESTIONS

Name Ethan O Date 2/23/24

True-False

(T)	F	**1.** Regulations are enforced by authorities having jurisdiction.
(T)	F	**2.** Under UL guidance, states may develop and administer their own occupational safety and health plans, as long as they cover public-sector employees and meet or exceed the federal regulations.
T	(F)	**3.** When approved, standards are identified by the OSHA acronym, sponsor organization acronym, and publication number of the organization.
(T)	F	**4.** The International Organization for Standardization (ISO) is a nongovernmental, international organization that provides a worldwide forum for the standards-developing process.
(T)	F	**5.** The CSA Group® is a Canadian government agency that tests products to verify conformance to national codes and standards.

Multiple Choice

C **1.** A(n) ___ is a designated unit of government with legal power to develop, administer, interpret, and enforce regulations.

A. standards organization
B. trade association
C. authority having jurisdiction
D. technical society

A **2.** The ___ is a federal government regulatory agency that requires all employers to provide a safe working environment for their employees.

A. Occupational Safety and Health Administration (OSHA)
B. National Institute for Occupational Safety and Health (NIOSH)
C. National Fire Protection Association (NFPA)
D. CSA Group

D **3.** The ___ does not develop standards, but it acts as a national coordinator of standards-writing activities and an approval organization and clearinghouse for standards.

A. National Fire Protection Association (NFPA)
B. American Welding Society (AWS)
C. National Electrical Manufacturers Association (NEMA)
D. American National Standards Institute (ANSI)

B ______ 4. The ___ is a national trade association that provides information and develops standards concerning proper selection, rating, construction, testing, and performance of electrical equipment.

A. National Institute for Occupational Safety and Health (NIOSH)
B. National Electrical Manufacturers Association (NEMA)
C. National Fire Protection Association (NFPA)
D. Occupational Safety and Health Administration (OSHA)

B ______ 5. The ___ is an independent organization that tests products to verify conformance to national codes and standards.

A. National Fire Protection Association (NFPA)
B. UL
C. IEEE
D. American National Standards Institute (ANSI)

Completion

regulation ______ 1. A(n) ___ is a rule made mandatory by a federal, state, or local government.

NIOSH ______ 2. The ___ is a national organization that acts in conjunction with OSHA to develop and periodically revise recommended exposure limits for hazardous substances or conditions in the workplace.

Standard ______ 3. A(n) ___ is a collection of voluntary rules that are developed through consensus and related to a particular trade, industry, or environment.

technical society ______ 4. A(n) ___ is an organization of technical personnel united by a professional interest.

NFPA ______ 5. The ___ publishes the National Electrical Code® (NEC®), which is one of the most widely adopted and recognized standards in the world.

PERSONAL PROTECTIVE EQUIPMENT

SECTION **2.3**

REVIEW QUESTIONS

Name Ethan Ø Date 2/23/24

True-False

(T)	F	**1.** Personal protective equipment (PPE) is clothing and/or equipment worn by a worker to reduce the possibility of an injury.
(T)	F	**2.** Arc-rated clothing should be clearly identified by a manufacturer's label indicating the level of protection.
T	(F)	**3.** Some tasks may require specialized eye protection due to especially dangerous hazards, such as the radiant energy from welding.
T	(F)	**4.** Long-term respiratory hazards include dusty, toxic, or oxygen-deficient atmospheres that quickly cause injury or death.
T	(F)	**5.** A long-term exposure limit (LTEL) is an exposure limit that allows a higher exposure level than the TWA for up to 15 minutes, as long as the average exposure during the 8 hr shift is below the TWA limit.
(T)	F	**6.** Advanced respirators include cartridges with special filter materials that absorb chemicals as the outside air passes through them.
(T)	F	**7.** Proper fit is critical for safe respirator operation.
T	(F)	**8.** Hearing protection is required whenever workers are exposed to noise equal to or less than 65 decibels (dB), which is equivalent to the sound level of busy street traffic.
(T)	F	**9.** Neoprene gloves provide protection against alcohol and a variety of chemicals and acids.
(T)	F	**10.** A warning line is a high-visibility rope marking a controlled-access zone near the edge of a roof.

Multiple Choice

C **1.** Class ___ protective helmets protect against impact hazards and contact with voltages up to 2200 V.

A. C (conductive)
B. E (electrical)
C. G (general)
D. H (hazardous)

A ______ **2.** ___ are an eye protection device that fully encloses the eye area with a flexible frame.
- A. Goggles
- B. Safety glasses
- C. Face shields
- D. Welding masks

D ______ **3.** A ___ is an exposure limit based on constant exposure for 8 hr/day, 40 hr/week.
- A. threshold limit value (TLV)
- B. ceiling limit
- C. short-term exposure limit (STEL)
- D. time-weighted average (TWA)

B ______ **4.** Gloves made from wire mesh, leather, or ___ provide protection from cuts and burns.
- A. rubber
- B. canvas
- C. latex
- D. denim

C ______ **5.** Fall protection equipment is required when working in elevated spaces (over ___′ above the ground) or confined spaces.
- A. 2
- B. 4
- C. 6
- D. 8

Completion

Protected Clothing ______ **1.** ___ is clothing that provides protection from contact with sharp objects, hot equipment, and harmful materials.

Protective helmet ______ **2.** A(n) ___ is a head protection device that prevents head injury from impact, falling and flying objects, and electrical shock in the workplace.

face Shield ______ **3.** A(n) ___ is an eye and face protection device that covers the entire face with a clear or tinted plastic shield.

Permissible exposure limit ______ **4.** The ___ limit is the OSHA estimate of the average airborne concentration of a substance to which workers may be exposed day after day without adverse effect.

Air Purifying ______ **5.** A(n) ___ respirator is a respirator that removes airborne contaminants from the surrounding air before the air is inhaled by the worker.

Safety Shoes ______ **6.** ___ are a foot protection device with reinforced steel toes that provide protection against injuries caused by compression and impact.

ear muffs ______ **7.** ___ are a device worn over the ears to reduce the level of noise reaching the eardrums.

full body harness ______ **8.** A(n) ___ is a fall-protection device that evenly distributes fall-arresting forces throughout the body to prevent further injury.

FIRE SAFETY

SECTION **2.4**

REVIEW QUESTIONS

Name Ethan **Date** 2/23/24

True-False

(T)	F	**1.** Fire extinguishers are selected based on the class of fire.
T	(F)	**2.** Smothering a fire with fire blankets or dry sand extinguishes a fire by removing heat.
(T)	F	**3.** The number and type of fire extinguishers required is determined by the AHJ and is based on how quickly a fire may spread, potential heat intensity, and accessibility to the fire.

Multiple Choice

B **1.** Further study of combustion and extinguishing technologies has revealed that a fourth critical element of a fire is ___.

A. carbon dioxide
B. a sustaining chain reaction
C. location
D. sparking

B **2.** Class ___ fires are fires that burn oil, gas, grease, paint, and other flammable liquids.

A. A
B. B
C. C
D. D

B **3.** Class ___ fires are fires caused by energized electrical equipment, such as motors or transformers.

A. B
B. C
C. D
D. K

Completion

fire safety plan **1.** A(n) ___ is a written document and public signage that establishes procedures that must be followed if a fire occurs.

A 2. Class ___ fires are fires that burn wood, paper, textiles, and other ordinary combustible materials.

D 3. Class ___ fires involve certain combustible metals, such as titanium, magnesium, sodium, and potassium.

ELECTRICAL SAFETY

SECTION 2.5

REVIEW QUESTIONS

Name Ethan V Date 2/23/24

True-False

(T) F 1. Electrical shock occurs when an individual becomes part of an electrical circuit, such as when touching an energized conductor.

T (F) 2. As current decreases, the severity of the electrical shock increases.

(T) F 3. An arc flash is an extremely high-temperature discharge produced by an electrical fault in the air.

T (F) 4. The amount of arcing that passes through a body depends on the voltage and resistance of the resulting circuit.

(T) F 5. A potential cause for arc flash and arc blast is improper test instrument use.

(T) F 6. The arc flash boundary distance and PPE requirements are based on the available incident energy of a circuit.

(T) F 7. The level of PPE required for work with energized electrical circuits is determined by the type of task, the voltage level and available fault current of the circuits involved, the proximity of the technician to live conductors, and other factors.

T (F) 8. Leather protectors are leather gloves worn under rubber insulating gloves to protect the hands and lower arms.

Multiple Choice

D 1. The severity of electrical shock depends on voltage, current, duration, ___, and other factors.

A. air temperature
B. body heat
C. exposure limit
D. frequency

C 2. Barriers such as insulation blankets, along with proper PPE, must be used to protect against hazards from nearby ___.

A. approach boundaries
B. dead circuits
C. energized circuits
D. arc flashes

B **3.** Only qualified personnel are allowed to work on energized circuits of ___ V or higher.
- A. 25
- B. 50
- C. 75
- D. 100

A **4.** A(n) ___ is a document that requires personnel to plan their task procedures and any related hazards when working around energized electrical circuits.
- A. energized electrical work permit (EEWP)
- B. work order
- C. confined space permit
- D. logbook

A **5.** Rubber insulating gloves are classified by their ___.
- A. maximum use voltage
- B. thickness
- C. color
- D. resistance

Completion

Arcing **1.** ___ is the discharge of electric current across an air gap.

Arc blast **2.** A(n) ___ is an explosion that may accompany an arc-fault event.

Qualified **3.** A(n) ___ person is a trained person with skills and knowledge in the construction and operation of electrical equipment and the recognition and avoidance of related hazards.

Arc flash **4.** A(n) ___ boundary is the distance from live parts within which individuals could receive a second-degree burn due to an arc flash.

incident energy **5.** The ___ is the thermal energy of an arc fault at a distance.

limited approach **6.** The ___ boundary is the distance from an exposed energized conductor or circuit part at which a shock hazard exists and is the closest distance an unqualified person may approach.

arc rating **7.** A(n) ___ is a PPE rating for the maximum amount of thermal energy an article of PPE can resist before exposing the wearer to second-degree burns.

rubber insulating **8.** ___ gloves are gloves used to provide protection from electrical shock hazards by insulating the hands and lower arms from contact with energized conductors.

SPECIAL PROCEDURES

SECTION **2.6**

REVIEW QUESTIONS

Name Otto P Date 2/23/24

True-False

T (F) **1.** Lockout is the attachment of a danger tag to a source of power to indicate that the equipment may not be operated until the tag is removed.

T (F) **2.** A lockout/tagout shall only be removed by the individual who installed it, even in an emergency.

(T) F **3.** Oxygen deficiency in confined spaces is caused by the displacement of oxygen by leaking gases or vapors, combustion or oxidation processes, oxygen absorbed by the vessel or product stored, and/or consumption by bacteria.

(T) F **4.** A plant may develop its own confined space permit form, but OSHA requires that certain checklist items be included.

(T) F **5.** A GHS label contains much more information about a chemical than other hazard communication labels, but it still includes graphical information to help convey safety warnings quickly.

Multiple Choice

C **1.** A ___ does not prevent the startup of equipment, but it serves as a warning to operating and service personnel.

A. valve cover
B. lockout
C. tagout
D. lockout hasp

A **2.** Standby personnel, commonly called a(n) "___," must remain just outside a confined space while workers are inside.

A. hole watch
B. fire watch
C. atmospheric tester
D. entry supervisor

C 3. NFPA 704, *Standard System for the Identification of the Hazards of Materials for Emergency Response,* uses a ___-color diamond sign to display basic information about hazardous materials.
- A. two
- B. three
- C. four
- D. five

B 4. A(n) ___ hazard is the degree of susceptibility of a material to explode or release energy by itself or by exposure to certain conditions or substances.
- A. electrical
- B. instability/reactivity
- C. health
- D. flammability

D 5. An SDS must be presented in ___ and include certain information required by OSHA 29 CFR 1910.1200—Hazard Communication.
- A. four hazard categories
- B. multiple languages
- C. pictogram form
- D. a 16-section format

Completion

Lockout 1. ___ is the use of locks, chains, or other physical restraints to prevent the operation of specific equipment.

Confined Space 2. A(n) ___ is a space large enough that an employee can physically enter and perform assigned work, but has limited or restricted means for entry and exit, and is not designed for continuous employee occupancy.

explosive 3. The lower ___ limit is the lowest concentration (air-fuel mixture) at which a gas can ignite.

Permit required 4. A(n) ___ confined space is a confined space that has specific health and safety hazards capable of causing death or serious physical harm.

Safety data Sheet 5. A(n) ___ is a document about a chemical that contains hazard information, including required PPE, storage and handling instructions, and response in case of personal exposure, fire, or chemical spill.

INDUSTRIAL HYGIENE

SECTION **2.7**

REVIEW QUESTIONS

Name ____________________ Date ____________________

True-False

T F 1. Many hazards identified by an industrial hygiene program are reduced by maintaining clean and organized workspaces.

T F 2. Biological hazards are the cause of many safety problems and accidents, particularly during night shifts.

T F 3. The degree of risk from exposure to any given substance depends on the nature and potency of toxic effects and magnitude and duration of exposure.

T F 4. HIV is the most common infectious bloodborne hazard faced by workers.

T F 5. OSHA, along with state agencies, designs rules to protect workers from exposure to serious diseases while performing lifesaving services.

Multiple Choice

______ 1. ___ is the study of the effects of job-related tasks and work-area arrangement on the health of a worker.
- A. Safety engineering
- B. Accident prevention
- C. Industrial hygiene
- D. Ergonomics

______ 2. A(n) ___ hazard is a hazard caused by excessive levels of noise, vibration, illumination, temperature, or radiation.
- A. electrical
- B. physical
- C. biological
- D. chemical

______ 3. A(n) ___ hazard is a bacterium, virus, fungus, or other microorganism that can cause acute and chronic infections by entering the body directly or through breaks in the skin.
- A. biological
- B. chemical
- C. electrical
- D. safety

Completion

industrial hygiene **1.** ___ is the study of anticipating, recognizing, evaluating, and controlling workplace conditions that may cause worker injury or illness.

Chemical **2.** A(n) ___ hazard is a chemical in any form that has toxic effects when inhaled, absorbed, or ingested.

blood borne Pathogen **3.** A(n) ___ is a pathogenic (disease-causing) microorganism present in human blood that can infect and cause disease in individuals through exposure.

EMERGENCIES

SECTION **2.8**

REVIEW QUESTIONS

Name Ethan ______________________ **Date** 2/23/24 ______________

True-False

(T) F **1.** Emergencies in an industrial facility have significant potential for facility and property damage, business disruption, and personnel injuries.

(T) F **2.** Planning for possible natural disasters may include developing procedures for notifying personnel of emergencies; closing the facility to minimize damage, secure hazardous materials, and protect against property loss; and maintaining minimum operations during and after the event.

(T) F **3.** When evaluating an emergency situation, the individual taking charge must obtain information from witnesses, assess the condition of the victim, and communicate with the victim.

Multiple Choice

___A___ **1.** ___ are responsible for performing additional duties that can save lives and prevent serious damage during an emergency.

A. Emergency coordinators
B. Service technicians
C. Supervisors
D. Qualified persons

___C___ **2.** A(n) ___ is a document that details facts about an injury-causing accident in a facility.

A. logbook
B. data log
C. incident report
D. safety data sheet

___C___ **3.** Incidents that result in the death of an employee or in-patient hospitalization of three or more employees must be reported to the ___ within 8 hr of the incident.

A. National Institute for Occupational Safety and Health (NIOSH)
B. National Fire Protection Association (NFPA)
C. Occupational Safety and Health Administration (OSHA)
D. authority having jurisdiction (AHJ)

Completion

Emergency Plan **1.** A(n) ___ is a document that details procedures, exit routes, and assembly areas for facility personnel in the event of an emergency.

explosion **2.** A(n) ___ is a blast that is caused by a bomb or the ignition of existing flammable material.

first aid **3.** ___ is emergency care or treatment given to an injured or ill individual before professional medical assistance is available.

WORKPLACE SAFETY

Chapter 2

ACTIVITIES

Name ______________________________ **Date** ______________

Fall Protection

______ **1.** Locking snap

______ **2.** D-ring

______ **3.** Lanyard

______ **4.** Shock absorber

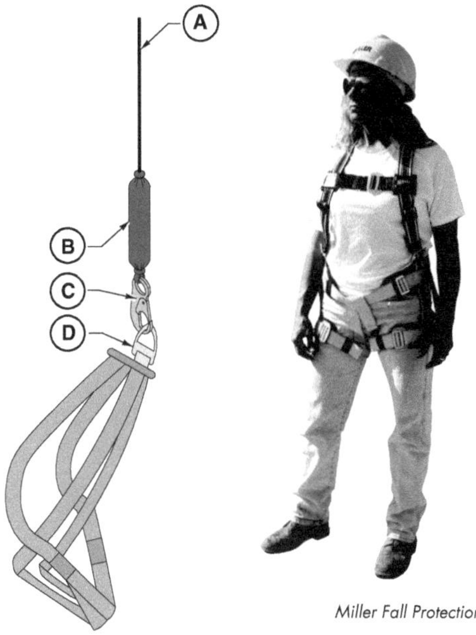

Miller Fall Protection

Protective Helmets

______ **1.** Class G

______ **2.** Class E

______ **3.** Class C

______ **4.** Type I

______ **5.** Type II

A. Special service, no voltage protection

B. Utility service, high voltage protection

C. General service, limited voltage protection

D. Impact from top and sides

E. Impact from top

Respirator Types

_______________ **1.** Reusable full-face respirator

_______________ **2.** Reusable half-mask respirator

_______________ **3.** Disposable particulate mask

_______________ **4.** Powered air-purifying respirator (PAPR)

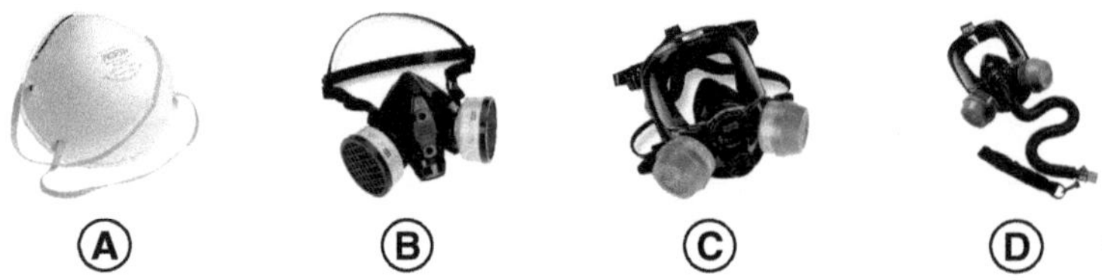

North Safety Products

Effects of Electric Current

_______________ **1.** Less than 8 mA

_______________ **2.** 8 mA to 15 mA

_______________ **3.** 16 mA to 20 mA

_______________ **4.** Over 20 mA

A. Sensation of shock but probably not painful

B. Painful shock; removal from contact point by natural reflexes

C. Painful shock; may be frozen or locked to point of electrical contact until circuit is deenergized

D. Causes severe muscular contractions, paralysis of breathing, heart convulsions

Fire Extinguisher Classes

__________ **1.** Ordinary combustibles

__________ **2.** Combustible metals

__________ **3.** Flammable liquids

__________ **4.** Electrical equipment

__________ **5.** Cooking oils

Gloves

__________ **1.** Rubber insulating

__________ **2.** Neoprene

__________ **3.** Latex

__________ **4.** Wire mesh

A. Protects against cuts

B. Provides protection against chemicals and acids

C. Provides maximum touch sensitivity

D. Provides protection against electrical hazards

NFPA 704 Diamond

__________ **1.** Fire hazard (red)

__________ **2.** Health hazard (blue)

__________ **3.** Specific hazard

__________ **4.** Instability/reactivity (yellow)

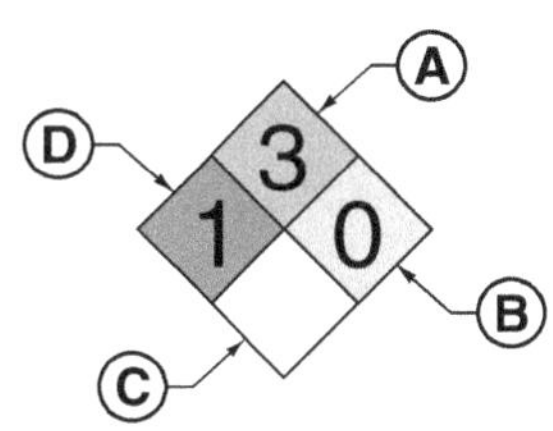

Activity: Safety Practices

Answer the questions based on the given situations.

1. What could be the atmospheric problem if a technician develops a headache and slight nausea after working in a boiler room for several hours?

2. What could be the atmospheric problem if a technician becomes suddenly weak and feels slightly disoriented while working inside a boiler?

3. Why must eye protection be worn when working with chemicals?

4. Why should an electric tool not be used while standing on a damp or wet floor?

5. What two items of personal protective equipment could make this situation safer?

Activity: Fire Safety

Answer the questions based on the given situations.

1. What might happen if a stream of water were used on this type of fire?

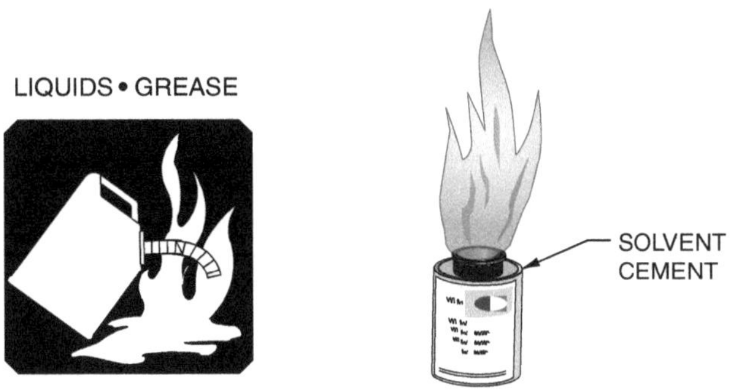

B FLAMMABLE LIQUIDS

2. What types of extinguishing agents cannot be used on this type of fire?

C ELECTRICAL EQUIPMENT

3. How are Class K and Class B fires similar?

4. List three types of fire extinguishing equipment.

Activity: Electrical Safety

Answer the questions based on the given situations.

1. Could a technician be seriously injured or killed from working on a 120 V lighting circuit with 100 W bulbs?

2. Why is it advisable to use one hand when working on electrical equipment?

3. If a technician temporarily stops work on a deenergized circuit, what must the technician do before resuming any work on the circuit?

4. What are potential causes of arc flashes or arc blasts?

Activity: Lockouts and Tagouts

Electrical power to an area of a manufacturing plant is locked out.

1. How many technicians are working on this circuit?

2. When can the circuit be energized?

A pneumatically actuated robotic arm in a manufacturing plant must be repaired.

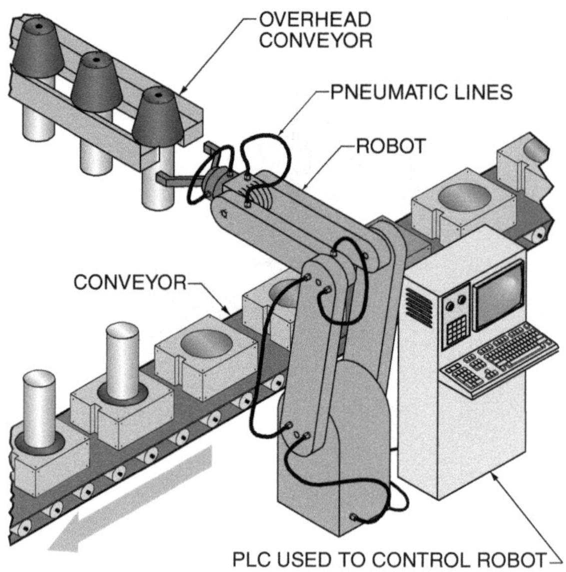

3. List two sources of energy that must be locked out for a technician to work on the arm.

The overhead conveyor is to be shut down for a major repair.

4. List two hazards that should be locked out and/or secured.

An air compressor feeding a receiver tank is to be shut down for a major repair.

5. List two places where lockouts should be applied.

6. What other energy source should be released?

Activity: Hazard Communication

Answer the questions based on the chemical label.

1. What is the name of the chemical?

2. What is the most significant hazard associated with this chemical?

3. What types of PPE are recommended when handling this chemical?

Describe the health, fire, instability/reactivity, and specific hazards detailed on each label.

4. Health hazard

5. Fire hazard

6. Instability/reactivity

7. Specific hazard

8. Health hazard

9. Fire hazard

10. Instability/reactivity

11. Specific hazard

PRINTS

SECTION 3.1

REVIEW QUESTIONS

Name ______________________________ **Date** ______________________

True-False

T F **1.** Prints that are not drawn to scale should be indicated with "Full Size" or the abbreviation "FS".

T F **2.** The use of symbols is preferred to abbreviations on prints whenever possible.

T F **3.** Zones are referenced by alphabetical and numerical entries in the margins.

T F **4.** A legend is also known as a bill of materials.

T F **5.** Typically, the context of a drawing clarifies the meaning of the abbreviation.

Multiple Choice

________________ **1.** Architectural prints are denoted by a capital ___.
- A. A
- B. M
- C. P
- D. S

________________ **2.** Title blocks are located on the ___ of a drawing sheet.
- A. top
- B. left side
- C. reverse
- D. right side

________________ **3.** A(n) ___ is a simple graphic that represents a thing or concept.
- A. abbreviation
- B. note
- C. symbol
- D. legend

________________ **4.** A revision is identified in a drawing on a print by a revision symbol, which is a(n) ___ inside of a triangle, circle, or square.
- A. number or letter
- B. asterisk
- C. letter "R"
- D. exclamation point

______ **5.** A(n) ___ note is a note that applies to the entire print set that the note appears on.
A. sheet
B. general
C. architecture
D. global

Completion

______ **1.** A(n) ___ is a reproduction of a technical drawing and its associated technical information.

______ **2.** For a large set of prints, an abbreviation list may be included as part of a(n) ___.

______ **3.** A(n) ___ note is a note that applies to a specific item in the drawing that the note appears with.

______ **4.** A(n) ___ is a list of information in a concise and organized format.

______ **5.** Prints for small parts may be at full size, which is noted as ___.

OBJECT REPRESENTATION

SECTION **3.2**

REVIEW QUESTIONS

Name ______________________________ **Date** ____________________

True-False

T F **1.** Leader lines are slanted lines.

T F **2.** A ghost line indicates additional shape, orientation, or position information about an object.

T F **3.** A view is chosen based on which view conveys the most information in the most efficient way.

T F **4.** For a cylindrical object, the half section appears as one-eighth of the object.

T F **5.** Overall dimensions are placed the farthest from the view.

Multiple Choice

________________ **1.** A(n) ___ line identifies the surface of a part created by a cutting-plane line.
- A. object
- B. extension
- C. section
- D. break

________________ **2.** A(n) ___ view is a pictorial view with the three axes 120° apart.
- A. sectional
- B. isometric
- C. equilateral
- D. triangular

________________ **3.** If there is more than one sectional view, ___ are typically used to match the cutting plane to the associated sectional view.
- A. letters
- B. numbers
- C. symbols
- D. leader lines

________________ **4.** A(n) ___ section is a sectional view in which the cutting-plane line is bent at an angle, and the section view is revolved as if flattened.
- A. half
- B. broken-out
- C. offset
- D. aligned

_______________ **5.** A(n) ___ dimension is indicated by the placement of parenthesis around the dimension.

A. reference
B. overall
C. intermediate
D. phantom

Completion

_______________ **1.** A(n) ___ line is a line that represents shapes that cannot be seen.

_______________ **2.** A(n) ___ line is a line that connects a dimension note or specification with a particular feature of a drawn object.

_______________ **3.** A(n) ___ view is a drawing that shows only one surface of an object, which is represented in true view.

_______________ **4.** A(n) ___ drawing is a type of drawing that provides supplemental information for small or complex features.

_______________ **5.** ___ is a permissible deviation from a given value or dimension.

MAINTENANCE PRINTS

SECTION 3.3

REVIEW QUESTIONS

Name ______________________________ **Date** ______________________

True-False

T F **1.** Elevations may detail outdoor structures that require maintenance, including culverts, parking lots, lights, and fences.

T F **2.** A floor plan is view from approximately 5′ above floor level.

T F **3.** Exploded view isometric drawings show the relationships of individual components with dashed lines to connect the various components in order.

T F **4.** Instructional drawings are typically drawn as exploded views.

T F **5.** A programmable logic controller (PLC) program diagram is a graphical representation of instructions saved into a PLC.

T F **6.** An operational diagram provides specific details on how a system operates.

Multiple Choice

______________ **1.** A ___ plan is a plan that depicts an entire property, including buildings and other structures drawn in their proper locations on the property.

A. utility
B. site
C. foundation
D. floor

______________ **2.** A(n) ___ is an orthographic drawing of a structure's vertical surfaces.

A. elevation
B. foundation plan
C. floor plan
D. assembly drawing

______________ **3.** A(n) ___ is a drawing that shows the position of switches, terminal connections, components, and other features on a device.

A. mounting dimension
B. installation drawing
C. clearance requirement
D. location drawing

______________ **4.** ___ diagrams resemble a ladder, so they are also known as ladder diagrams.
A. Schematic
B. Block
C. Wiring
D. Line

______________ **5.** A ___ is a drawing that illustrates the operation of hydraulic and pneumatic systems.
A. fluid power diagram
B. piping diagram
C. piping and instrumentation diagram (P&ID)
D. sequence of operation

______________ **6.** A time schedule for a system can be found in the ___ and includes daily, weekly, monthly, and annual schedules.
A. detail drawing
B. general notes
C. sequence of operation
D. title block

Completion

______________ **1.** A(n) ___ is a two-dimensional drawing of an object as it is viewed from above.

______________ **2.** A(n) ___ drawing is a drawing that shows as closely as possible the way individual parts or components are placed together to produce the finished equipment or arrangement.

______________ **3.** A(n) ___ dimension is a dimension that indicates the minimum space required to install a piece of equipment.

______________ **4.** A schematic diagram is a detailed drawing showing all the components of a system and their interconnections using ___ and symbols.

______________ **5.** A(n) ___ is a narrative that details the operation of a system and its interaction with other systems in a logical order.

______________ **6.** A(n) ___ is a drawing that shows the piping and control equipment in an industrial or commercial process.

ADDITIONAL DOCUMENTATION

SECTION **3.4**

REVIEW QUESTIONS

Name ________________________________ **Date** ____________________

True-False

T F **1.** Equipment specifications are developed and published by equipment manufacturers.

T F **2.** Safety concerns, explanations of operating controls, and alarm signal explanations are noted throughout an equipment specification.

T F **3.** Over time, most equipment develops tendencies or repeating problems.

Multiple Choice

______________ **1.** A(n) ___ is a single piece of information, typically numerical, related to a piece of equipment's characteristics.

A. project specification
B. equipment specification
C. operator's manual
D. service bulletin

______________ **2.** Updates to an operator's manual may be distributed as ___.

A. email blasts
B. forum posts
C. recalls
D. service bulletins

______________ **3.** A ___ is a collection of various documents that are assembled by a technician for easy reference and related to a certain system or machine.

A. working manual
B. troubleshooting report
C. technician's journal
D. service bulletin

Completion

______________ **1.** A(n) ___ is a document included with a set of prints that contains additional information that cannot be illustrated on a print.

______________ **2.** A(n) ___ is a record of a specific problem that occurs in a particular piece of equipment, along with its symptoms, causes, and repair procedures.

______________ **3.** A working manual is limited to ___ piece(s) of equipment.

______________ **4.** A(n) ___ manual is a document that contains instructions for the safe and efficient installation, operation, troubleshooting, and repair of equipment.

PRINTREADING

Chapter 3

ACTIVITIES

Name ______________________________ **Date** ____________________

Scales

A print is drawn at a 1:1 scale.

__________ **1.** If the actual feature is 3″ in length, its representation in the drawing is ___″.

__________ **2.** If the actual feature is 27 mm in length, its representation in the drawing is ___ mm.

__________ **3.** If a drawing dimension is 5.25″, the actual feature is ___″.

A print is drawn at a ½″ = 1′ scale.

__________ **4.** If the actual feature is 7.0′ in length, its representation in the drawing is ___″.

__________ **5.** If a drawing dimension is 10.5″, the actual feature is ___′.

__________ **6.** If a drawing dimension is 7.25″, the actual feature is ___′.

A print is drawn at a 1:20 scale.

__________ **7.** If the drawing dimension is 20 mm, the actual feature is ___ mm.

__________ **8.** If the drawing dimension is 6.25 cm, the actual feature is ___ cm.

__________ **9.** If the actual feature is 50 cm, its representation in the drawing is ___ cm.

__________ **10.** If the actual feature is 6.5 m, its representation in the drawing is ___ cm. (Note the units.)

Line Types

______ **1.** Object line

______ **2.** Hidden line

______ **3.** Centerline

______ **4.** Dimension line

______ **5.** Extension line

______ **6.** Leader line

______ **7.** Cutting-plane line

______ **8.** Section lines

______ **9.** Break line

______ **10.** Phantom line

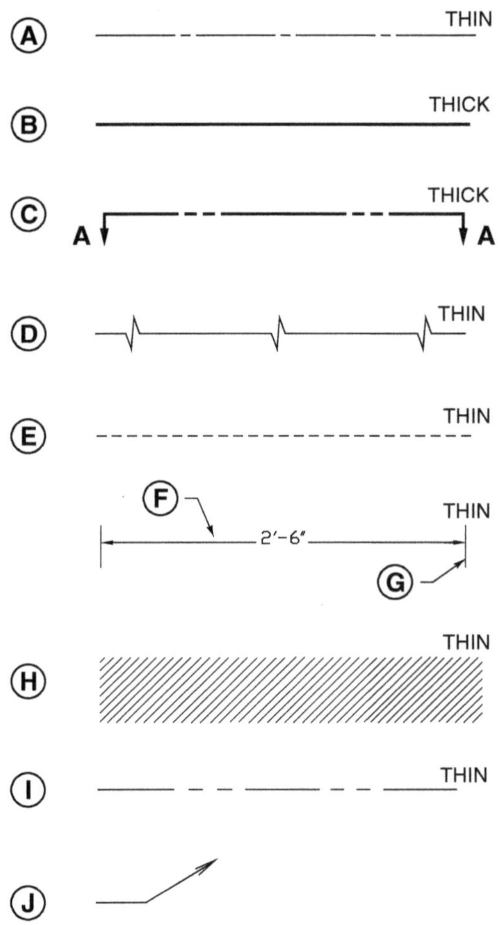

Sectional Views

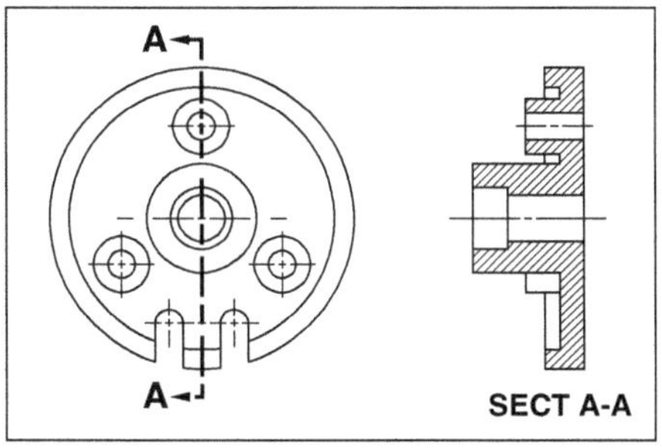

Ⓐ

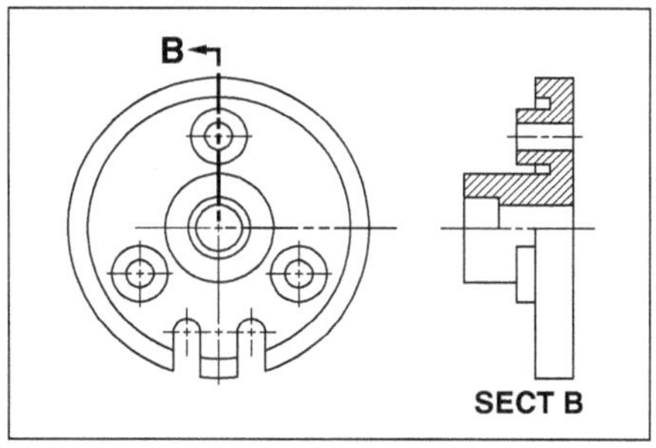

Ⓑ

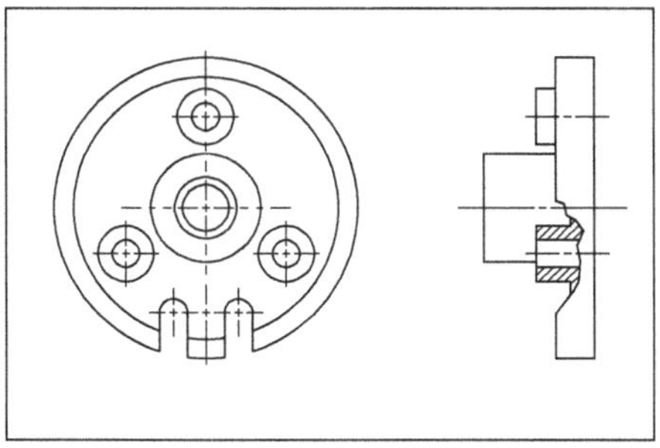

Ⓒ

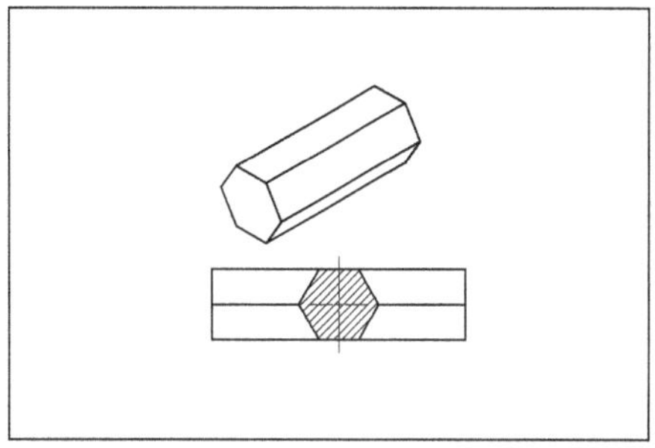

Ⓓ

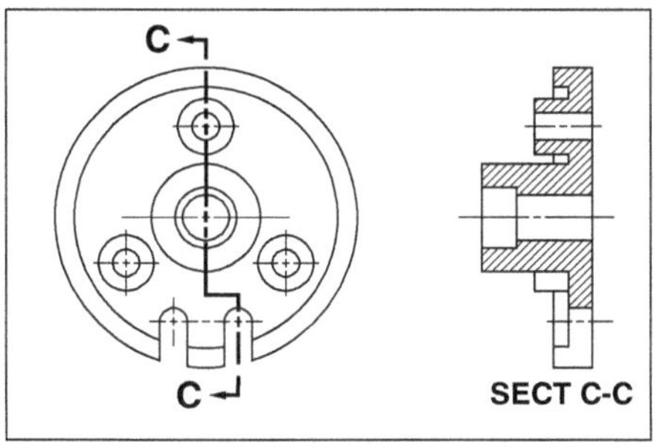

Ⓔ

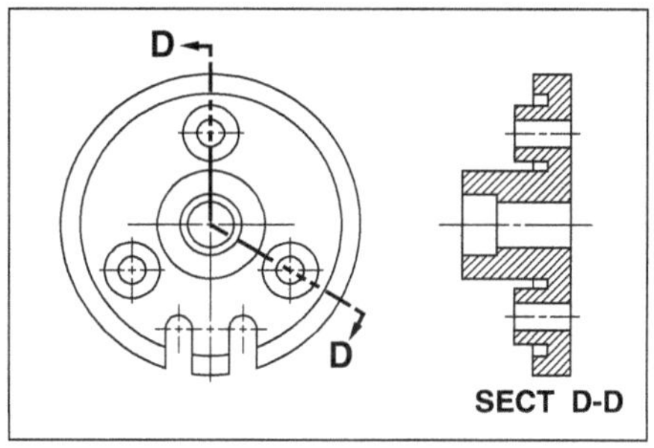

Ⓕ

______________ **1.** Aligned section

______________ **2.** Revolved section

______________ **3.** Full section

______________ **4.** Offset section

______________ **5.** Half section

______________ **6.** Broken-out section

Installation Drawings

______________________ **1.** Mounting dimensions

______________________ **2.** Feature information

______________________ **3.** Clearance requirements

______________________ **4.** Outline dimensions

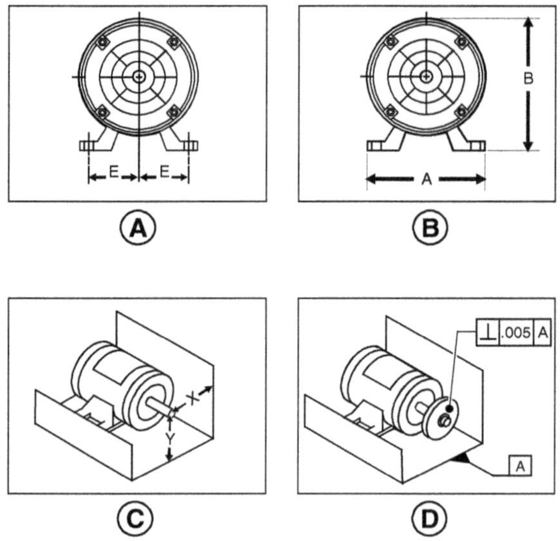

Maintenance Prints

Determine the best type of print to consult in the following situations.

______________________ **1.** Locating underground utilities for the entire property

______________________ **2.** Troubleshooting the operation of an electrical control circuit

______________________ **3.** Understanding how individual parts come together to form finished equipment

______________________ **4.** Visualizing the appearance of a building from outside

______________________ **5.** Determining the amount of space required for a new piece of equipment

______________________ **6.** Troubleshooting the operation of a PLC

______________________ **7.** Comparing the input and response characteristics of a sensor

______________________ **8.** Tracing the flow of oil in a hydraulic system

______________________ **9.** Learning how to operate or adjust a piece of equipment

______________________ **10.** Studying the arrangement of rooms and built-in features for a building

A. Installation drawing
B. Fluid power diagram
C. Line diagram
D. Instructional drawing
E. Elevation
F. Floor plan
G. Site plan
H. Assembly drawing
I. PLC program diagram
J. Operational diagram

Activity: Title Blocks

Answer the questions based on the given title block.

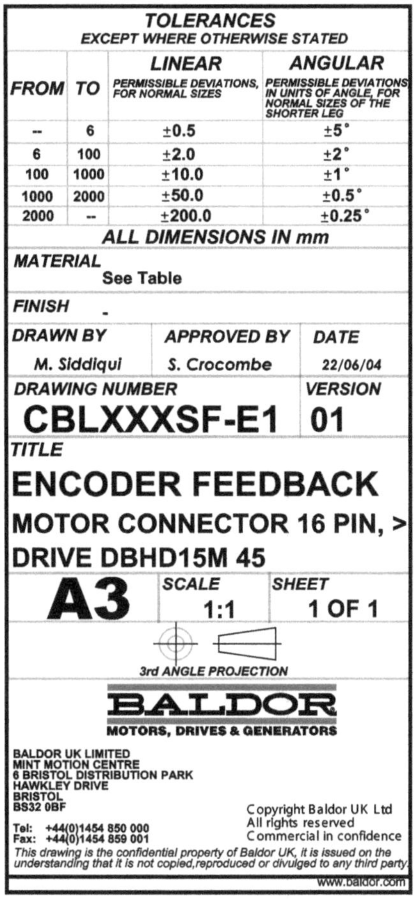

TOLERANCES EXCEPT WHERE OTHERWISE STATED			
FROM	TO	LINEAR PERMISSIBLE DEVIATIONS, FOR NORMAL SIZES	ANGULAR PERMISSIBLE DEVIATIONS IN UNITS OF ANGLE, FOR NORMAL SIZES OF THE SHORTER LEG
--	6	±0.5	±5°
6	100	±2.0	±2°
100	1000	±10.0	±1°
1000	2000	±50.0	±0.5°
2000	--	±200.0	±0.25°

ALL DIMENSIONS IN mm

MATERIAL See Table

FINISH -

DRAWN BY	APPROVED BY	DATE
M. Siddiqui	S. Crocombe	22/06/04

DRAWING NUMBER	VERSION
CBLXXXSF-E1	01

TITLE

ENCODER FEEDBACK
MOTOR CONNECTOR 16 PIN, >
DRIVE DBHD15M 45

A3	SCALE 1:1	SHEET 1 OF 1

3rd ANGLE PROJECTION

BALDOR
MOTORS, DRIVES & GENERATORS

BALDOR UK LIMITED
MINT MOTION CENTRE
6 BRISTOL DISTRIBUTION PARK
HAWKLEY DRIVE
BRISTOL
BS32 0BF

Tel: +44(0)1454 850 000
Fax: +44(0)1454 859 001

Copyright Baldor UK Ltd
All rights reserved
Commercial in confidence

This drawing is the confidential property of Baldor UK, it is issued on the understanding that it is not copied,reproduced or divulged to any third party

www.baldor.com

1. What is the name of this part?

2. Who is the manufacturer of this part?

3. What units are used for dimensions in this drawing?

4. What is the scale of this drawing?

5. At this scale, a 10.50 mm part feature would be drawn at what size?

6. What is the linear tolerance for a 10.50 mm feature?

7. Therefore, what is the range of acceptable sizes for this feature?

Activity: Printreading

Answer the following questions based on the provided print.

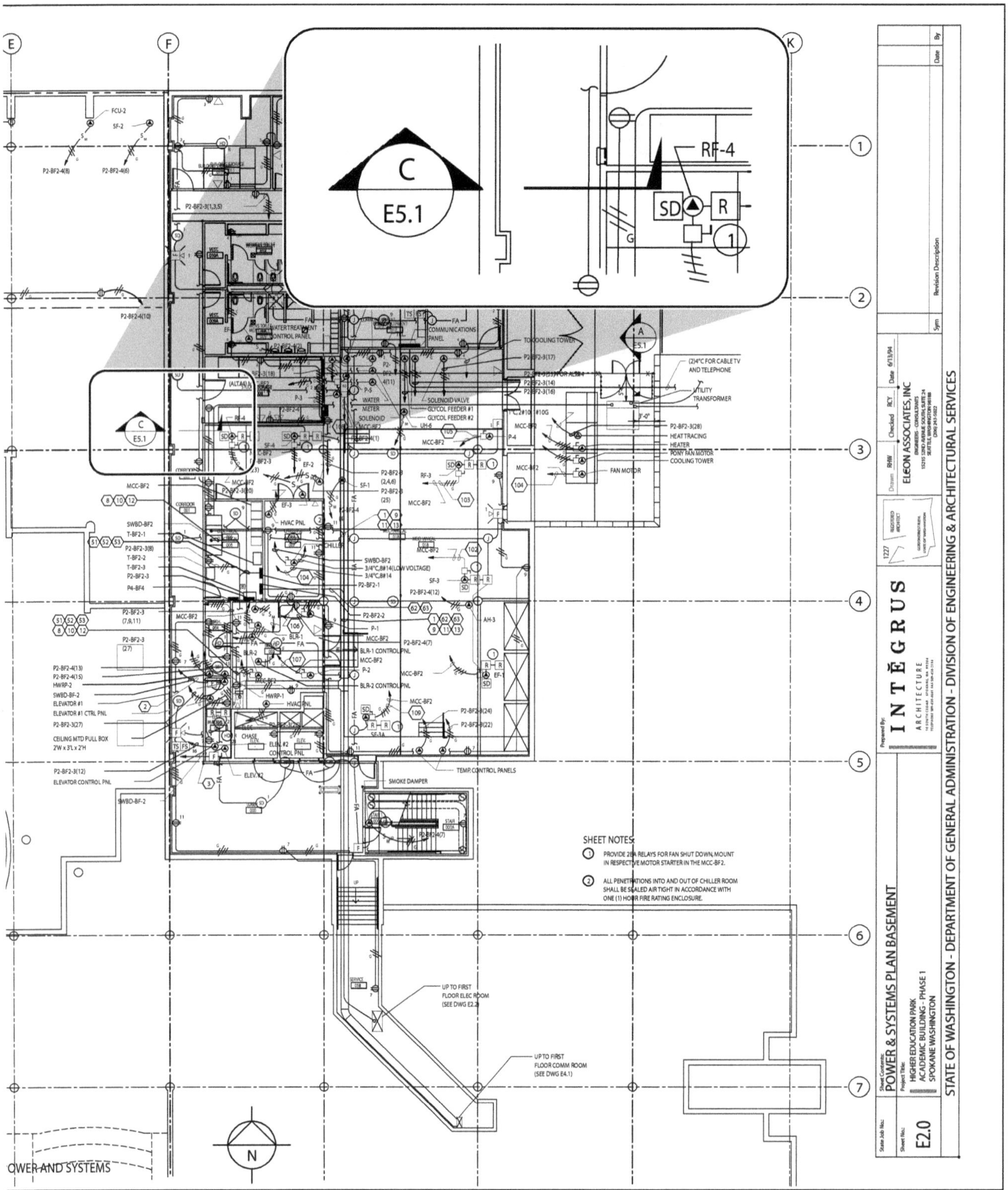

1. What type of print is this?

2. What part of the building does this print show?

3. What is the print division?

4. How many sheet notes are included?

5. What is the zone for Sectional View C?

6. In which direction is Sectional View C directed?

7. What sheet are the sectional views on?

Activity: Detail Drawings

Answer the following questions based on the provided print.

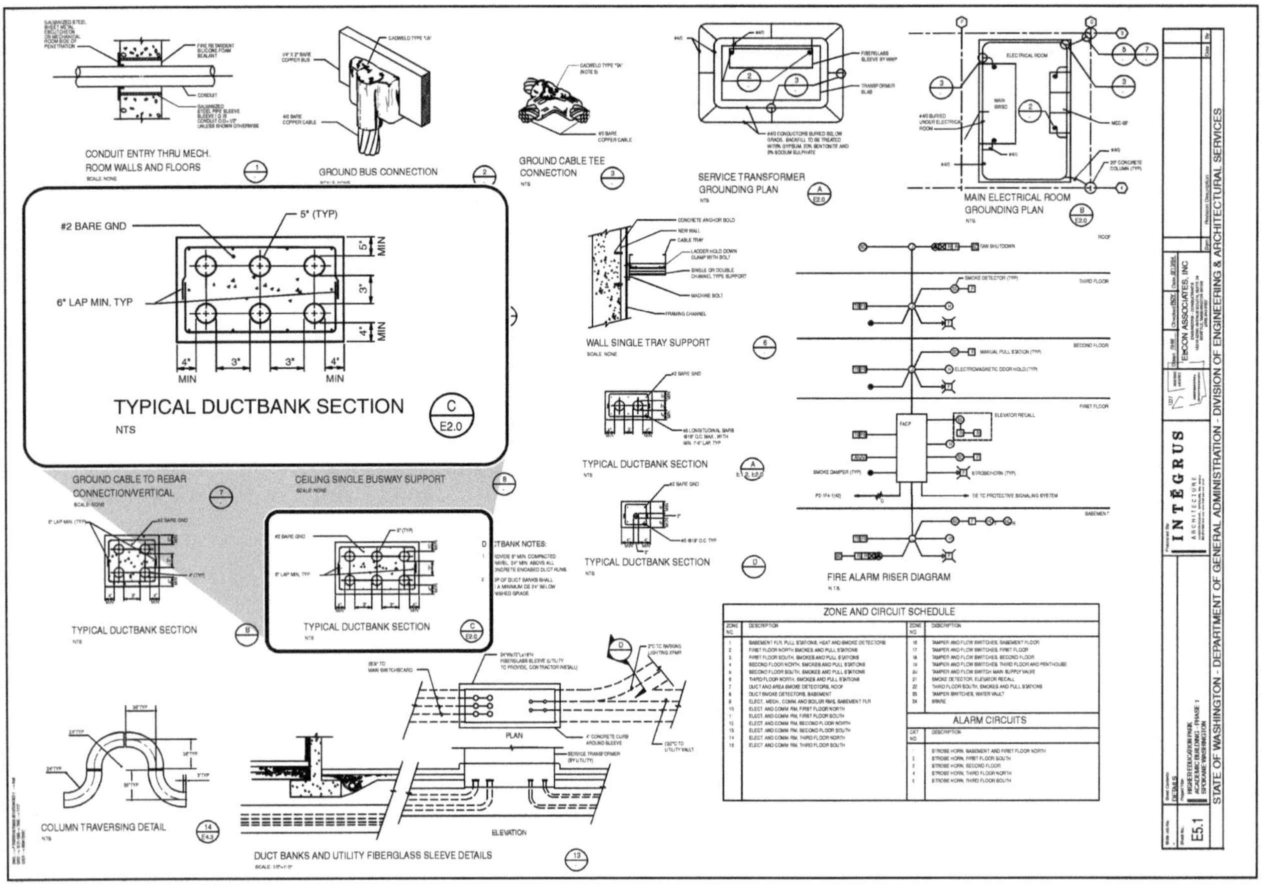

Integrus Architecture

1. What types of drawings or other documentation are included on this sheet?

2. What is the name of Sectional View C?

3. What is the scale of Sectional View C?

4. Which sheet references Section View C?

5. What material is to be used in the center of the ductbank?

MECHANICS

SECTION **4.1**

REVIEW QUESTIONS

Name ________________________________ **Date** ____________________

True-False

T F **1.** A compressed spring has kinetic energy because of its condition.

T F **2.** Force acting on an object does not always result in motion.

T F **3.** A polished metal surface has a lower coefficient of friction than rough sandpaper.

T F **4.** A lever is a simple machine consisting of a cylinder rotating freely on its axis, which uses a belt, rope, or chain to change the direction of an applied force.

T F **5.** A screwdriver is an example of a wedge.

Multiple Choice

______ **1.** Force can be measured in ___.
- A. kilograms
- B. pounds
- C. horsepower
- D. pound-feet

______ **2.** ___ is the number of unit squares equal to the surface of an object.
- A. Work
- B. Torque
- C. Area
- D. Pressure

______ **3.** ___ is the force that resists the relative motion of two objects sliding against each other.
- A. Friction
- B. Torque
- C. Power
- D. Inertia

______ **4.** A ___-class lever is a lever that has the effort located between the fulcrum and the resistance.
- A. first
- B. second
- C. third
- D. fourth

D 5. A(n) ___ is a simple machine that converts force applied to its blunt end into force that is perpendicular to its sloped surface.
   A. inclined plane
   B. pulley
   C. screw
   D. wedge

Completion

Pressure 1. ___ is force per unit area.

Power 2. ___ is the rate of doing work or using energy.

Efficiency 3. ___ is a measure of a device's useful output energy compared to its input energy.

Kinetic 4. ___ energy is energy of motion.

Potential 5. ___ energy is stored energy a body has due to its position, chemical state, or condition.

Work 6. ___ is the movement of an object by a force to a specific distance.

torque 7. ___ is rotational force.

inertia 8. ___ is the tendency of a physical body to persist in its state of rest or uniform motion until acted upon by an external force.

HEAT

SECTION **4.2**

REVIEW QUESTIONS

Name ____________________ **Date** ____________

True-False

T F **1.** The state of a substance depends on the intensity of the motion of the molecules.

T F **2.** Heat added to a substance causes a decrease in molecular motion.

T F **3.** Heat transfer rates increase as the temperature difference between two substances increases.

Multiple Choice

________ **1.** ___ is heat transfer that occurs within a fluid when currents circulate between warm and cool regions.

- A. Radiation
- B. Conduction
- C. Convection
- D. Induction

________ **2.** On the Celsius scale, ___°C is the freezing point.

- A. 0
- B. 32
- C. 100
- D. 212

________ **3.** Heat produced on earth by light from the sun is a form of ___.

- A. convection
- B. conduction
- C. induction
- D. radiation

Completion

________ **1.** ___ is a form of energy identified by temperature difference or a change of state.

________ **2.** A(n) ___ is the quantity of heat required to raise the temperature of 1 lb of water 1°F.

________ **3.** ___ is the measurement of the intensity of heat.

SECTION **4.3**

REVIEW QUESTIONS

Name ________Ethan________ Date ______________

True-False

(T) F 1. Torsion is stress caused by forces acting in opposite twisting motions.

(T) F 2. Reduction is the loss or removal of oxygen from a material.

T (F) 3. The color test is a metal identification test that identifies metals by the appearance of the spark emitted from contact with a grinding wheel.

Multiple Choice

____B____ 1. ___ deformation is the ability of a stressed material to return to its original size and shape after being unloaded.
- A. Plastic
- B. Elastic
- C. Temporary
- D. Strain

____A____ 2. High-___ materials resist stress from loads, deform easily, and fracture gradually.
- A. ductile
- B. brittle
- C. oxidation
- D. magnetic

____C____ 3. The ___ test is a metal identification test where a metal sample is broken to check for ductility and grain size.
- A. file
- B. chip
- C. fracture
- D. spark

Completion

____mechanical____ 1. A(n) ___ load is an external force applied to a body that causes stress in a material.

____thermal____ 2. ___ expansion is the change in volume of a material in relation to temperature.

____stress____ 3. ___ is the internal effect of an external force applied to a solid material.

mulleability 4. ___ is the ability of a material to be deformed by compression without developing defects.

Physical 5. ___ properties include thermal, electrical, and other properties of a material.

FASTENING

SECTION 4.4

REVIEW QUESTIONS

Name ______ Ethan ______ Date ______

True-False

(T)	F	**1.** The shielded metal arc welding (SMAW) process uses electricity to generate the heat necessary for melting workpieces.
(T)	F	**2.** Adhesive bonding is useful for joining dissimilar metals, plastics, and composites in manufacturing and repair operations.
T	(F)	**3.** Brazing and soldering produce joints that are stronger than welded joints.
(T)	F	**4.** Low-viscosity adhesives are liquid in form and flow readily into small spaces.
(T)	F	**5.** Flat washers are used under the head of a screw or bolt or under a nut to spread a load over a greater area.

Multiple Choice

___B___ **1.** Bolts and screws are classified into grades based on their ___.
- A. length and shape
- B. tensile strength and yield strength
- C. head style
- D. thread lock coating

___D___ **2.** A(n) ___ is a tool for cutting internal threads.
- A. washer
- B. fastener
- C. extractor
- D. tap

___B___ **3.** A ___ is a cylindrical nonthreaded fastener that is placed into a hole to secure the position of two or more parts.
- A. bolt
- B. pin
- C. stud
- D. key

___C___ **4.** ___ keys are used for transmitting heavy torques in shafts and hubs that are reversed frequently and subject to vibration.
- A. Alternating
- B. Parallel
- C. Taper
- D. Woodruff

B 5. ___ is an arc welding process that uses a separate shielding gas and a continuous wire electrode.

A. Oxyfuel welding (OFW)
B. Gas metal arc welding (GMAW)
C. Shielded metal arc welding (SMAW)
D. Gas tungsten arc welding (GTAW)

A 6. A(n) ___ adhesive is a one- or two-part adhesive or sealant that cures by evaporation or catalyst.

A. polysulfide
B. cyanoacrylate
C. polyurethane
D. anaerobic

Completion

Woodruff 1. A(n) ___ key is shaped like a half-circle.

Welding 2. ___ is a joining process that fuses materials by heating them to melting temperature.

brazing 3. ___ is a joining process that joins parts by heating a filler metal to temperatures greater than 840°F but less than the melting point of the base metal.

Viscosity 4. ___ is the measure of the resistance of a fluid to flow.

helix 5. A(n) ___ is the curve traced on a cylinder or cone by a spiral.

SEALING AND COATING

SECTION **4.5**

REVIEW QUESTIONS

Name ______________________ **Date** ______________________

True-Falses

T F **1.** Sealing and coating procedures do not usually require shutdown, removal, and repair/replacement of the unit.

T F **2.** Patches can be permanent or temporary.

T F **3.** Sealing is the application of a surfacing material that has hardness properties to reduce damage from wear.

Multiple Choice

______________ **1.** Exposed metal parts of lathes and other precision equipment are coated with a ___ to prevent corrosion of the metal before the equipment is shipped.
- A. greaselike sealant
- B. gritty paint
- C. fine, clean sand
- D. primer

______________ **2.** A cleaning tank is used for cleaning parts in ___.
- A. nonflammable solution
- B. flammable solvents
- C. saline
- D. soapy water

______________ **3.** ___ are used to seal close-fitting pieces of metal such as parts of an engine or pipe flanges.
- A. Patches
- B. Sealants
- C. Spot welds
- D. Gaskets

Completion

______________ **1.** A(n) ___ is a rigid or semirigid pliable material placed between mating surfaces to prevent gas or liquid leakage.

______________ **2.** A(n) ___ is a product used to seal, fill voids, and waterproof parts.

______________ **3.** ___ is the process of applying a layer of material by welding or brazing to obtain the desired dimensions or surface properties.

REVIEW QUESTIONS

Name ______________________ Date ______________________

True-False

T F **1.** A hammer should never be used to strike another hammer.

T F **2.** A grinder is a common hand tool.

T F **3.** Stand to one side when starting and using a grinder.

Multiple Choice

______ **1.** Small work should be secured ___.
- A. by wrapping tape around the end farthest from the work
- B. using temporary sealants
- C. with nails
- D. in a clamp

______ **2.** Hand tools used for fastening include ___.
- A. wrenches
- B. grinders
- C. air chisels
- D. drill presses

______ **3.** Do not use an electric tool on ___.
- A. painted workpieces
- B. workpieces in a confined space
- C. wet or damp floors
- D. broken or splintered workpieces

Completion

______ **1.** Point cutting tools ___ the body during use.

______ **2.** Power tools are powered electrically, ___, or hydraulically.

______ **3.** When using ___ tools, use tool components specifically designed for compressed air service.

SERVICE AND REPAIR PRINCIPLES

Chapter 4

ACTIVITIES

Name ______________________________ **Date** ____________________

Torque

__________ **1.** A 30 lb force at 2′ develops ___ lb-ft of torque.

__________ **2.** A 45 lb force at 1.5′ develops ___ lb-ft of torque.

__________ **3.** A 10 lb force at 1′ develops ___ lb-ft of torque.

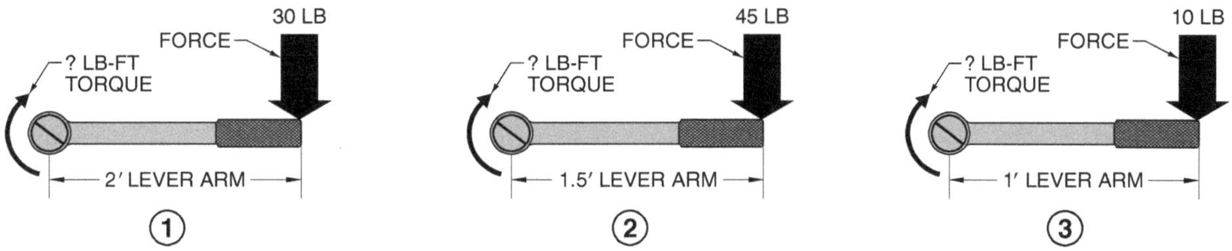

Lever Classes

__________ **1.** First class

__________ **2.** Second class

__________ **3.** Third class

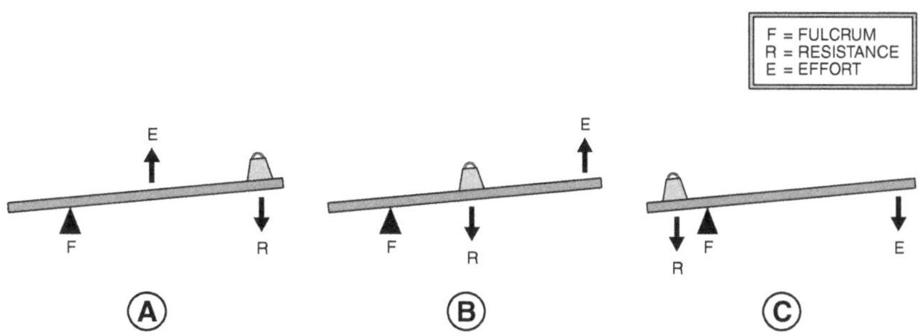

Horsepower

______________ **1.** A(n) ___ HP engine is required in order to lift 4950 lb to a height of 2′ in 3 sec.

______________ **2.** A(n) ___ HP engine is required in order to lift 49,500 lb to a height of 1′ in 1 min.

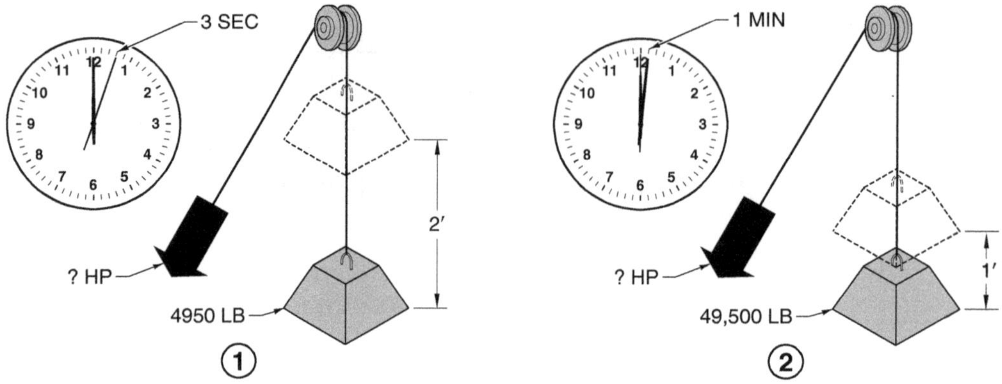

Stress

______________ **1.** Compression

______________ **2.** Bending

______________ **3.** Shear

______________ **4.** Tension

______________ **5.** Torsion

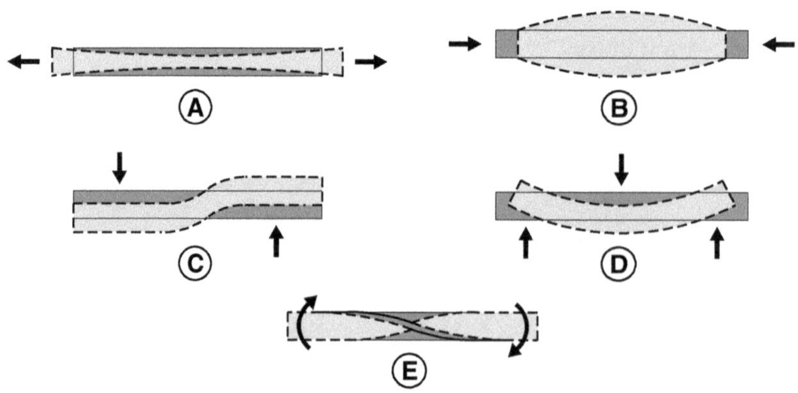

Temperature Conversion

_______________ **1.** The temperature at A is equivalent to ___°C.

_______________ **2.** The temperature at B is equivalent to ___°F.

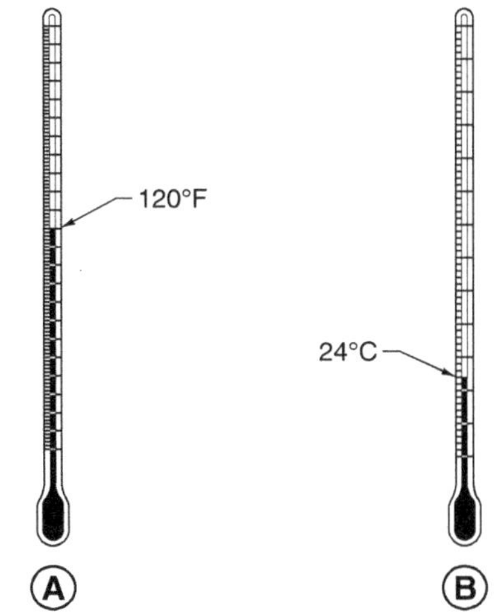

Washers

_______________ **1.** Tooth lock, external

_______________ **2.** Tooth lock, internal

_______________ **3.** Tooth lock, internal-external

_______________ **4.** Tooth lock, countersunk external

_______________ **5.** Flat

_______________ **6.** Spring lock

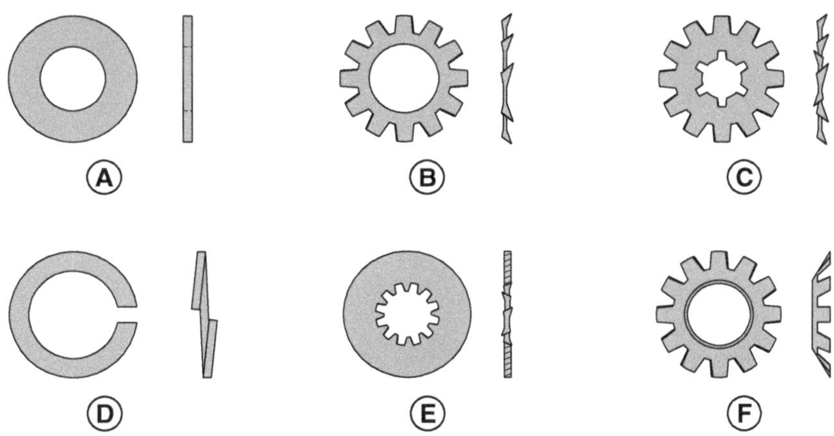

Screw Thread

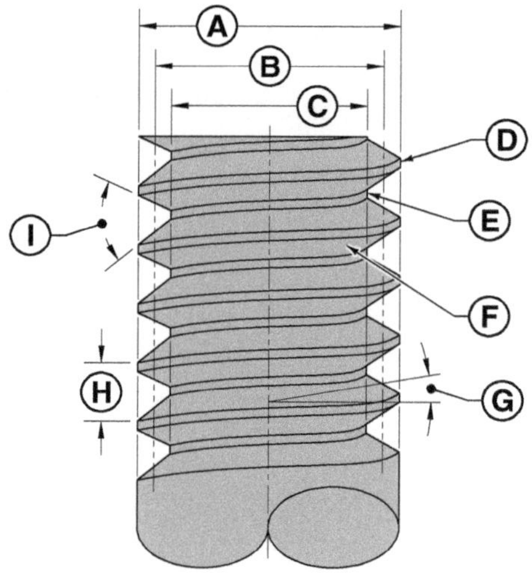

_______________ 1. Crest

_______________ 2. Helix angle

_______________ 3. Pitch

_______________ 4. Thread angle

_______________ 5. Flank

_______________ 6. Root

_______________ 7. Major diameter

_______________ 8. Pitch diameter

_______________ 9. Minor diameter

Pins

_______________ **1.** Grooved, type A

_______________ **2.** Grooved, type B

_______________ **3.** Grooved, type C

_______________ **4.** Cotter, extended prong, square cut type

_______________ **5.** Cotter, lock type

_______________ **6.** Straight

_______________ **7.** Dowel

_______________ **8.** Clevis

_______________ **9.** Slotted spring, enclosed slot

_______________ **10.** Slotted spring, straight slot

_______________ **11.** Spirally coiled

_______________ **12.** Taper

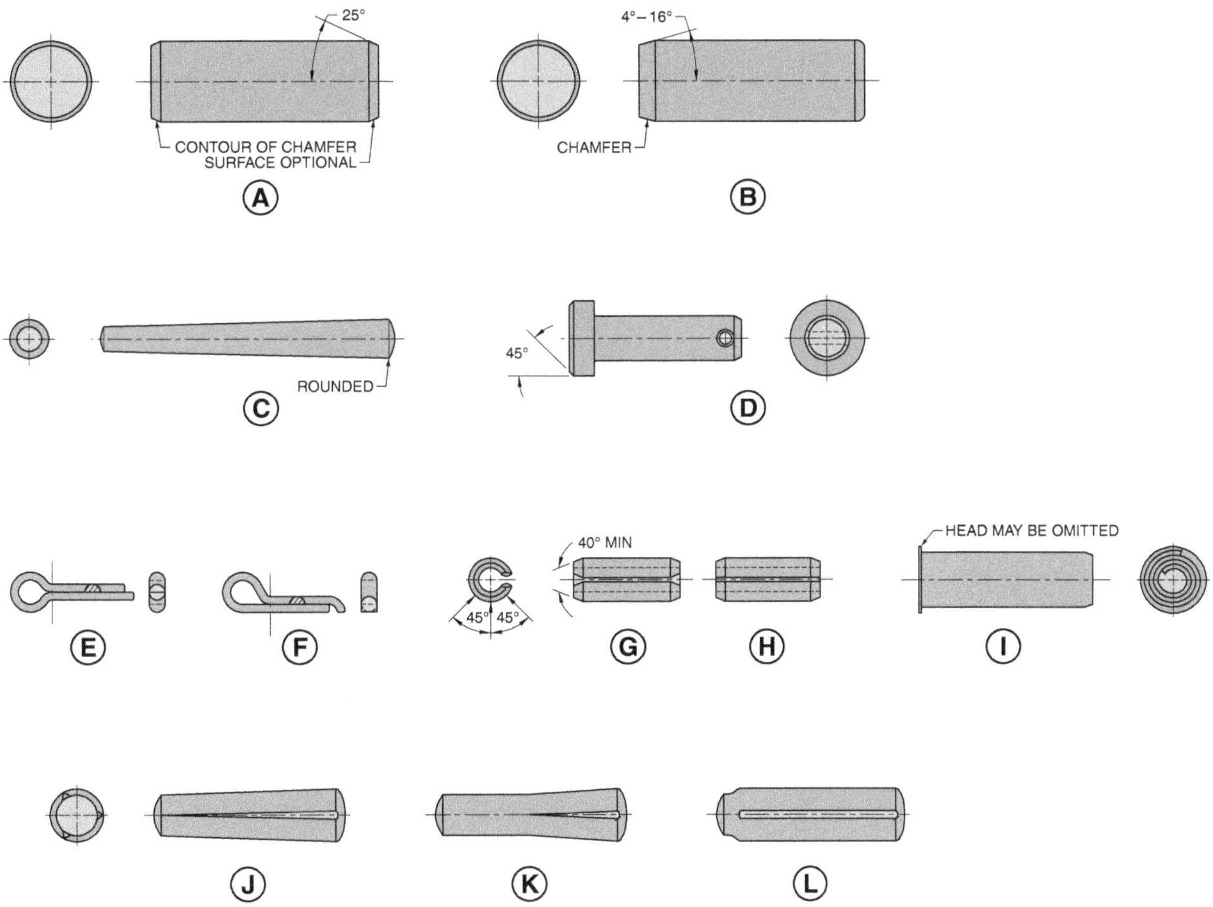

Activity: Energy and Forces

A technician is struggling to lift a weight, but it is not moving.

1. Is any work being done in this situation?

2. Is energy being expended?

A safety valve is held closed by a strong spring and can be hand-actuated by a long lever.

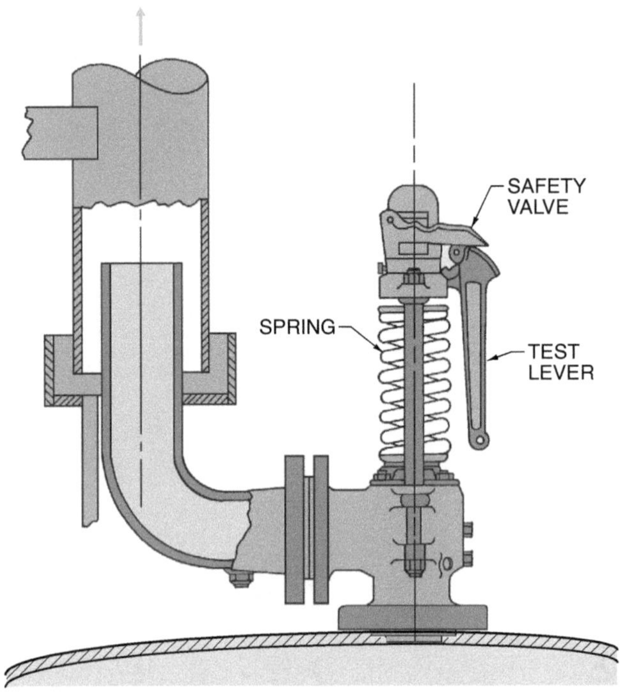

3. What type of energy does a spring represent when it is compressed?

4. Why must safety glasses be worn when working around compressed springs?

5. Why is the test lever designed to be long?

Two conveyors are used to transport products throughout a warehouse.

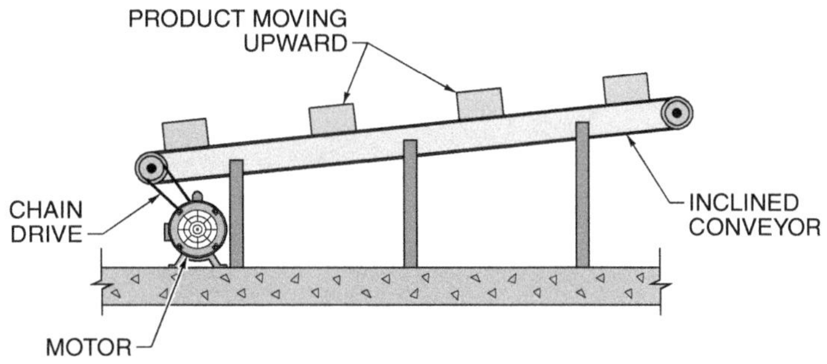

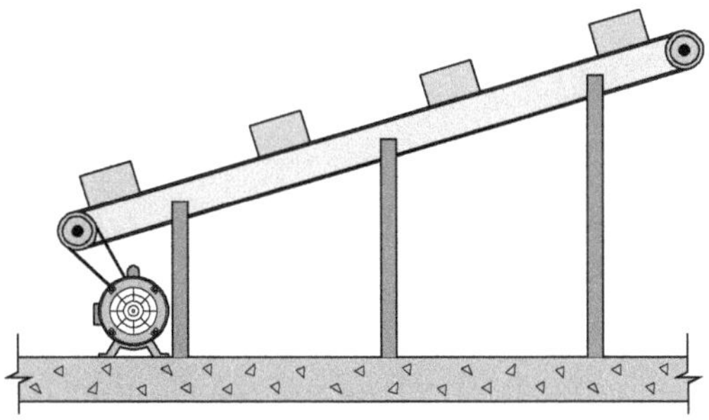

6. The conveyors are examples of which type of simple machine?

7. Which conveyor is doing more work?

Activity: Heat Transfer

Energy from the sun heats a multistory building.

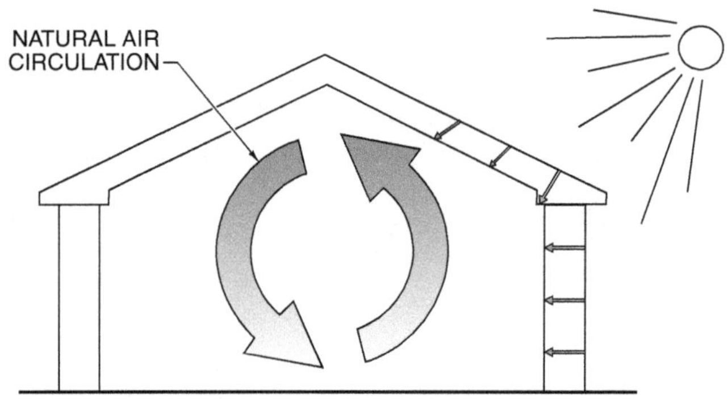

1. The sun's rays represent what type of heat energy?

2. What type of energy transfer occurs through the walls of the building?

3. Why might the ground floor feel cooler than the top floor of the building?

Activity: Fasteners and Mechanical Properties

Determine the type of mechanical stresses or forces on the following fasteners.

1. What type of stress causes a rivet to flatten when hit by a hammer?

2. What type of stress does a nail experience when it is pulled out by a hammer?

3. What type of force keeps a taper pin securely in a tightly fitted hole?

4. What type of stress will cause a cotter pin to break?

5. What type of stress is exerted when a stud or screw head is broken off?

6. What type of force is used to tighten a screw?

7. Explain the differences between welding, brazing, and soldering.

8. List at least five characteristics that may need to be considered when selecting an adhesive.

Activity: Tools

Explain the purpose of the following precautions.

1. Why should power tools be unplugged when they are being moved?

2. Why should operators stand to one side when starting and using a grinder?

3. Why should tools be pointed away from the body when used?

4. Why should lint-free rags be used when lubricating equipment and when working on hydraulic equipment?

5. Why should a brittle nail not be hit with a hammer?

MECHANICAL DRIVE TYPES

SECTION **5.1**

REVIEW QUESTIONS

Name ______________________________ **Date** ____________________

True-False

T F **1.** There is no slippage between the drive and driven parts of the machine in a nonsynchronous drive system.

T F **2.** A miter gear is a bevel gear used at right angles to transmit power at a 1:1 ratio.

T F **3.** A master link is the center link of a silent chain that has no teeth.

T F **4.** Any two external gears that mate rotate in opposite directions.

T F **5.** A fractional-horsepower (FHP) belt is a V-belt designed for heavy-duty applications.

Multiple Choice

______________ **1.** A(n) ___ is a motor, engine, or other rotary power source.

A. gearbox
B. impeller
C. drive system
D. prime mover

______________ **2.** ___ gears are the most common gears used in industry and are used where gears are mounted on parallel shafts.

A. Spur
B. Helical
C. Bevel
D. Worm

______________ **3.** A ___ is a steel plate used to connect two roller chain pins or two bushings.

A. guide link
B. side bar
C. silent chain
D. master link

______________ **4.** A belt drive is a ___ mechanical drive system that uses a belt and pulleys to transfer power between two shafts.

A. silent
B. synchronous
C. nonsynchronous
D. right-angle

______ **5.** The gear diameter determines the ___ of teeth on a gear.
A. number
B. type
C. size
D. shape

______ **6.** A cam ___ is a machine part that contacts the cam and moves in a designated path.
A. ear
B. lobe
C. tracer
D. follower

Completion

______ **1.** ___ is the distance between corresponding points on an adjacent pair of evenly spaced projections.

______ **2.** A(n) ___ is a flexible loop that connects two or more pulleys for transferring rotational power.

______ **3.** Belt ___ is the slight backward shifting of a belt relative to the drive pulley due to change in tension along the length of the belt.

______ **4.** A(n) ___ is a sealed container that has an input shaft and an output shaft and houses at least one set of mating gears.

______ **5.** A(n) ___ is a machine part that converts rotary motion into a custom, repeating sequence of linear or rotary motions by using an irregular external or internal surface.

______ **6.** A(n) ___ is a series of interconnected links that form a loop.

______ **7.** ___ are the most frequently used belts for belt drive systems.

______ **8.** A(n) ___ sprocket is a sprocket that has no hub.

LUBRICANTS

SECTION **5.2**

REVIEW QUESTIONS

Name ______________________ **Date** ______________

True-False

T F **1.** Friction and wear occur simultaneously.

T F **2.** Solid lubricants cannot be used in extreme operating conditions.

T F **3.** Lubricants may be liquid, semisolid, or solid.

T F **4.** The grease dropping point is the temperature at which the thickening agent of a grease turns to a solid.

T F **5.** Lubricants reduce the operating temperatures of machine parts by transferring heat from the metal surfaces to the lubricant.

Multiple Choice

______________ **1.** ___ indicates the expected thickness of the lubricant film separating the bearing surfaces.

A. Rank
B. Rating
C. Grade
D. Viscosity

______________ **2.** A(n) ___ is a chemical added to a primary substance in relatively small quantities to enhance a desired characteristic.

A. additive
B. contaminant
C. preservative
D. ingredient

______________ **3.** A ___ should be used when checking lubricant levels.

A. filter
B. lint-free cloth
C. cleansing wipe
D. spare rag or towel

Completion

______________ **1.** A(n) ___ is a substance that separates moving (bearing) surfaces to reduce the friction and wear between them.

______________ **2.** ___ is the forming of metal oxides as a result of metal and oxygen combining.

______________ **3.** Common lubricating ___ is not suitable for operating temperatures above 325°F.

______________ **4.** ___ is the measure of the resistance of a fluid's molecules to move past each other.

______________ **5.** ___ friction is friction that occurs when one surface moves across another or both surfaces move in opposite directions.

ANCILLARY COMPONENTS

SECTION **5.3**

REVIEW QUESTIONS

Name ______________________________ **Date** ____________________

True-False

T F **1.** A radial load is a load applied parallel to the rotating shaft.

T F **2.** A spider is a cross-shaped member having eight ends providing pivot points.

T F **3.** Elastomeric couplings are normally maintenance-free.

T F **4.** A cage is an antifriction bearing component used to maintain the position and alignment of the rolling elements.

T F **5.** An axial load is a load applied perpendicular to the rotating shaft.

Multiple Choice

______________ **1.** An oil seal ___ is an oil seal component that contacts the moving part of the equipment to prevent material from passing by the oil seal.
- A. film
- B. spacer
- C. lip
- D. case

______________ **2.** A ___ coupling is a mechanical-flexing coupling that consists of two identical hubs with external gear teeth and a sleeve or sleeves with mating internal gear teeth.
- A. universal
- B. block-and-jaw
- C. grid
- D. gear

______________ **3.** A(n) ___ coupling is a material-flexing coupling that includes two flanges separated by a floating member to accommodate shaft misalignment.
- A. diaphragm
- B. spring
- C. elastomeric
- D. flexible disk-ring

______ **4.** A(n) ___ bearing is a bearing that contains moving elements that provide a low-friction support surface for rotating or sliding surfaces.
A. friction
B. antifriction
C. wear
D. motion

______ **5.** A(n) ___ coupling is a device that joins two shafts that are precisely aligned.
A. rigid
B. flexible
C. chain
D. elastomeric

Completion

______ **1.** A(n) ___ is a machine component used to reduce friction and maintain clearance between stationary and moving parts.

______ **2.** A(n) ___ is a device used to connect two shafts, usually from a motor to a mechanical drive.

______ **3.** An oil ___ is a device used to prevent oil from leaking out through the hole in a housing where a shaft protrudes.

______ **4.** A(n) ___ metal is a metal that does not contain iron.

______ **5.** A(n) ___ bearing is an antifriction bearing that is designed to provide low-friction movement of a mechanical device that moves in a straight line.

MECHANICAL DRIVE PREDICTIVE MAINTENANCE

SECTION **5.4**

REVIEW QUESTIONS

Name ______________________________ **Date** ____________________

True-False

T F **1.** Manufacturer statistics, such as life span, may be used as a predictive maintenance tool.

T F **2.** High-frequency vibrations are caused by imbalance, misalignment, or a bent shaft.

T F **3.** Internal flaws or cracks do not affect sound waves as they pass through metals.

T F **4.** Temperature changes or extreme temperatures, either high or low, may indicate a condition that could cause problems in the future.

T F **5.** A wear particle analysis is the study of the size, frequency, shape, and composition of wear particles present in a lubricating oil.

Multiple Choice

______________ **1.** Vibration ___ is the speed of travel from an extreme limit of the vibration displacement to the neutral position.

A. frequency
B. acceleration
C. velocity
D. phase

______________ **2.** ___ analysis is a type of vibration analysis that compares current condition readings with baseline condition readings to detect a wide variety of changes.

A. Signature
B. Broadband
C. Narrowband
D. Ultrasonic

______________ **3.** A(n) ___ uses many temperature measurements within its field of view to build an infrared image of an area.

A. thermal imager
B. electronic thermometer
C. infrared thermometer
D. spectrometer

______________ **4.** A ___ is often used to easily draw oil from a machine without leaving a residue in the pump to contaminate future samples.
A. funnel
B. spectrometer
C. diaphragm coupling
D. vacuum sampling pump

______________ **5.** A(n) ___ vaporizes materials and records the resulting light.
A. thermal imager
B. spectrometer
C. electronic thermometer
D. infrared thermometer

Completion

______________ **1.** ___ is the monitoring of individual component vibration characteristics to analyze the component condition.

______________ **2.** ___ is the maximum range of motion from the upper limit to the lower limit of a vibration cycle.

______________ **3.** ___ analysis is analysis that uses high-frequency vibrations to detect anomalies or problems in mechanical systems.

______________ **4.** ___ is the use of temperature-indicating or temperature-measuring devices to quantify temperature or temperature changes.

______________ **5.** A(n) ___ is a predictive maintenance technique used to detect and analyze the presence of acids, dirt, water, fuel, and wear particles in oil.

MECHANICAL DRIVE PREVENTIVE MAINTENANCE

SECTION **5.5**

REVIEW QUESTIONS

Name ______________________________ **Date** ______________

True-False

T F **1.** Belt adjustments may be made at the head-end snub roller or the tail-end snub roller.

T F **2.** A sealed bearing serves as a conduit for oil from the oil cup to the bearing surface.

T F **3.** Hard foot is a condition that occurs when one or more machine feet do not make complete and solid contact with the base plate.

T F **4.** Power must always be off, and the power source must be locked out and tagged out before work is performed on mechanical drives.

T F **5.** Pullers may be mechanically or hydraulically operated.

Multiple Choice

______________ **1.** The lower portion of a gearbox is usually a ___.
- A. dipstick for checking oil level
- B. reservoir for lubricating internal gears
- C. bearing surface
- D. pressure cup

______________ **2.** A ___ is a device that measures the amount of deflection of a belt.
- A. tensiometer
- B. snub roller
- C. shim
- D. feeler gauge

______________ **3.** A ___ is a thin precision piece of material, usually metal, that provides an accurate spacing between two surfaces.
- A. coupling
- B. washer
- C. shim
- D. gasket

______________ **4.** Pullers are sized by force (in ___), spread, and reach.
- A. ounces
- B. pounds
- C. grams
- D. tons

Completion

______________ **1.** ___ is a slick polished surface caused by dirt and other debris being rubbed on the surface of a belt.

______________ **2.** A(n) ___ roller is a roller used to guide (track) a conveyor belt on conveyor pulleys.

______________ **3.** A(n) ___ is a device used to remove gears, pulleys, sprockets, bearings, and couplings from a shaft or housing.

______________ **4.** A(n) ___ cup is a pressurized grease reservoir that provides constant lubrication to a bearing.

______________ **5.** Proper belt tension should result in approximately ___″ of belt deflection for every inch of span between pulley centers.

MECHANICAL DRIVE TROUBLESHOOTING

SECTION **5.6**

REVIEW QUESTIONS

Name ____ **Date** ________________________

True-False

T F **1.** Unusual sounds, heat, vibration, motion, and smell are all indications that a problem exists.

T F **2.** New bearings are generally 75% to 85% filled with grease.

T F **3.** Electric current produces pitting, patterns of pitting, or fluted surfaces.

T F **4.** Corrosive wear is wear that occurs when a hard surface rubs against a softer surface.

T F **5.** Heating a bearing increases the size of the bearing, allowing it to slip over a shaft.

Multiple Choice

________________ **1.** Gear tooth failure can be classified as ___ deterioration or tooth breakage.
A. light
B. heavy
C. critical
D. surface

________________ **2.** ___ is plastic flow that occurs due to excessive loads or impact loading.
A. Heavy wear
B. Ridging
C. Rolling and peening
D. Rippling

________________ **3.** ___ is breakage caused by one large, sudden shock.
A. Fatigue
B. Overload
C. Ridging
D. Pitting

________________ **4.** Poor bearing ___ allow dirt and other contaminants to enter the bearing components.
A. bushings
B. rollers
C. cages
D. seals

______________ **5.** Oxide buildup is caused by the ___ on bearing surfaces.
- A. presence of moisture
- B. presence of dirt
- C. absence of lubrication
- D. absence of carbon

Completion

______________ **1.** ___ wear is wear that occurs when acid or moisture deteriorates a surface.

______________ **2.** Insufficient ___ may cause chain whip or the chain to climb the teeth of the sprocket.

______________ **3.** Common ___ symptoms include a distorted steel case, a damaged sealing spring, a seal cocked in its housing, and loose internal components.

______________ **4.** ___ is the change in the shape of a material caused by stress.

______________ **5.** ___ is the flaking off of a metal surface.

MECHANICAL DRIVES

Chapter 5

ACTIVITIES

Name ______________________________ **Date** ____________________

Belt Tension

______________ **1.** Drive pulley

______________ **2.** Driven pulley

______________ **3.** Tight side

______________ **4.** Slack side

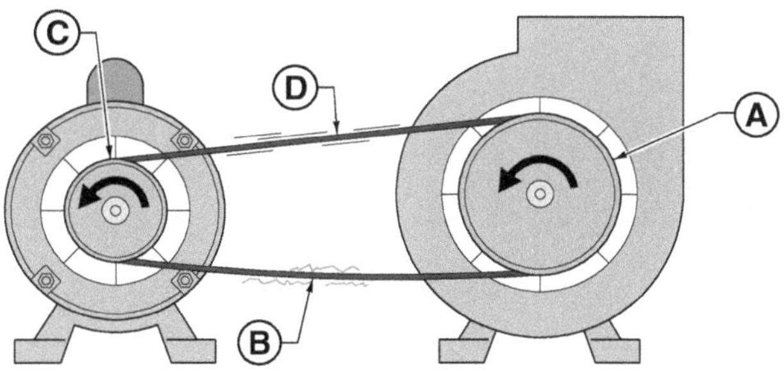

Removing and Installing Belts

List steps in sequential order.

______________ **1.**

______________ **2.**

______________ **3.**

______________ **4.**

______________ **5.**

______________ **6.**

______________ **7.**

______________ **8.**

______________ **9.**

______________ **10.**

A. Remove belt guard

B. Check belt tension

C. Remove old belt and check pulley wear

D. Turn off, lock out, and tag out machine

E. Move motor back using pry bar

F. Loosen belt tension

G. Move pulleys to required center-to-center distance and slide belt over pulleys

H. Replace guard and remove lockout/tagout

I. Select proper belt

J. Check pulley alignment

Roller Bearings

__________ **1.** Needle

__________ **2.** Cylindrical

__________ **3.** Spherical

__________ **4.** Tapered

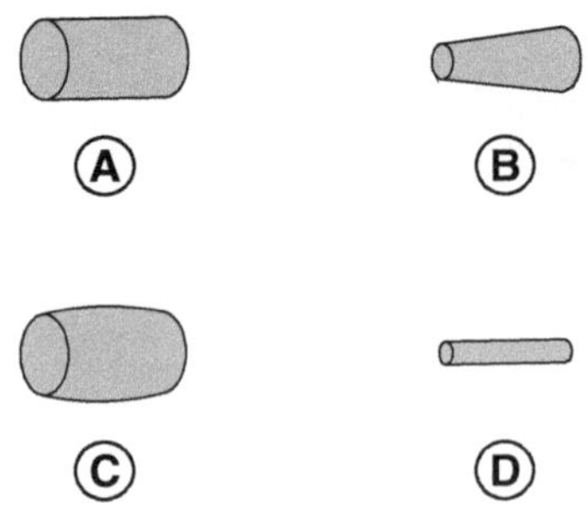

Belts

__________ **1.** Single V-belt

__________ **2.** Multiple V-belt

__________ **3.** Link-type V-belt

__________ **4.** Double-sided timing belt

__________ **5.** Flat belt

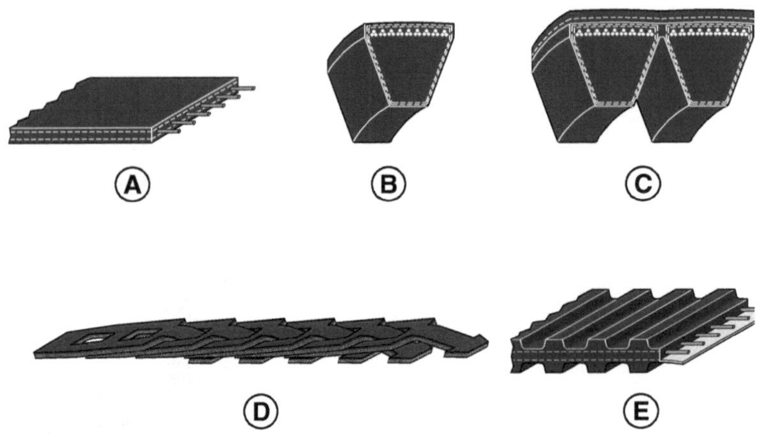

Roller Chain

__________ **1.** Pin

__________ **2.** Pitch

__________ **3.** Roller

__________ **4.** Roller width

__________ **5.** Roller diameter

__________ **6.** Code

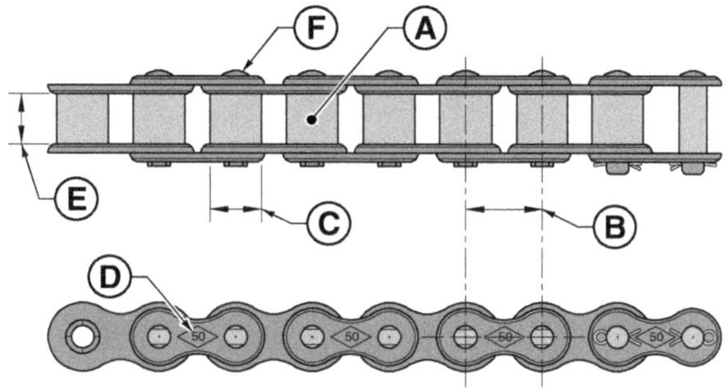

Belt Materials

__________ **1.** Natural rubber belt material has ___ oil resistance.

__________ **2.** Neoprene belt material has ___ tear resistance.

T F **3.** Natural rubber, neoprene, and polyurethane belt materials have excellent abrasion resistance.

T F **4.** Natural rubber belts operate in the same temperature range as polyurethane belts.

__________ **5.** A neoprene belt has ___ solvent resistance.

BELT MATERIALS							
Material	**Tensile Range***	**Temperature Range†**	**Solvent Resistance**	**Oil Resistance**	**Abrasion Resistance**	**Tear Resistance**	**Aging Due to Weather**
Natural rubber	500 to 3500	−60 to 175	Poor	Poor	Excellent	Excellent	Poor
Neoprene	500 to 3000	−50 to 185	Fair	Fair	Excellent	Good	Good
Polyurethane	500 to 6000	−30 to 175	Poor	Good	Excellent	Excellent	Excellent

Gears

______________________ **1.** Helical

______________________ **2.** Bevel

______________________ **3.** Worm

______________________ **4.** Miter

______________________ **5.** Herringbone

______________________ **6.** Spur, external

______________________ **7.** Spur, internal

______________________ **8.** Spur, rack and pinion

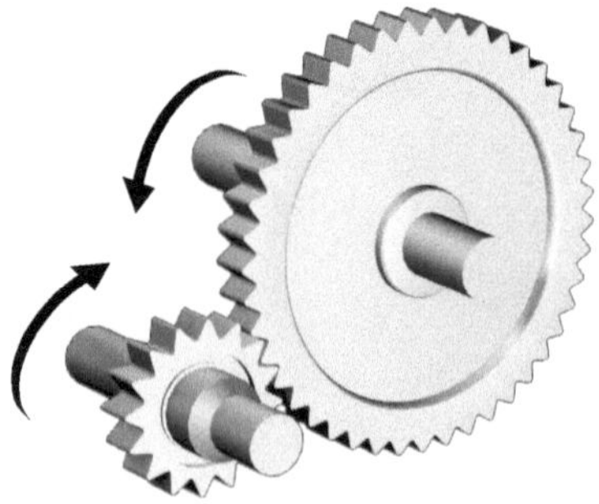

Ⓐ

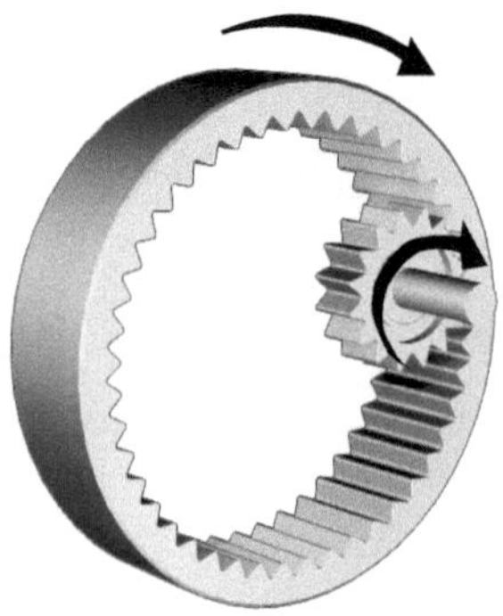

Ⓑ

Ⓒ

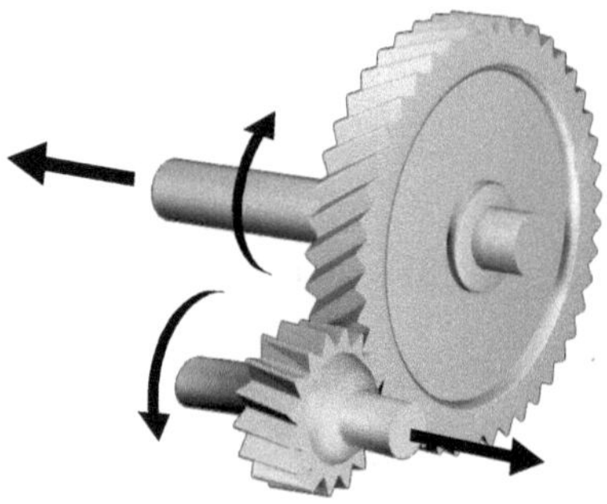

Ⓓ

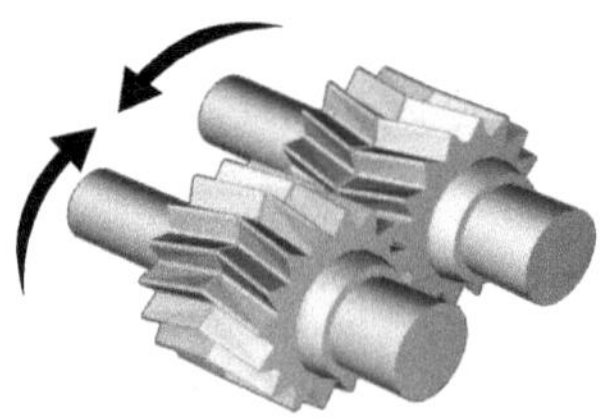

Ⓔ

Ⓕ

Ⓖ

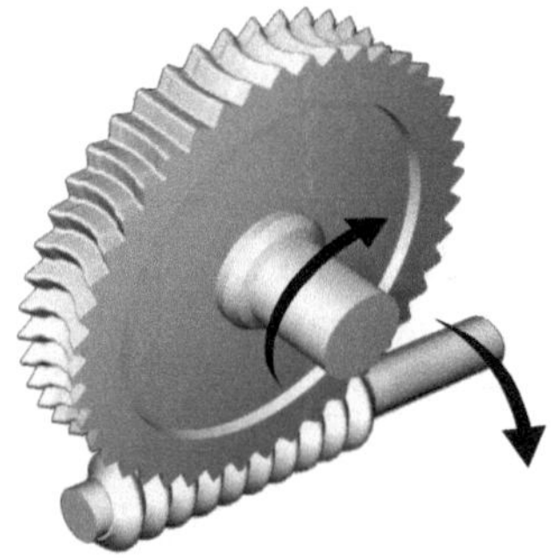

Ⓗ

Gear Rotation

______________ **1.** Gear B turns in a(n) ___ rotation.

______________ **2.** Gear C turns in a(n) ___ rotation.

T F **3.** Gear C and D turn at the same speed.

T F **4.** Gear B turns slower than Gear A.

______________ **5.** Gear D turns in a(n) ___ rotation.

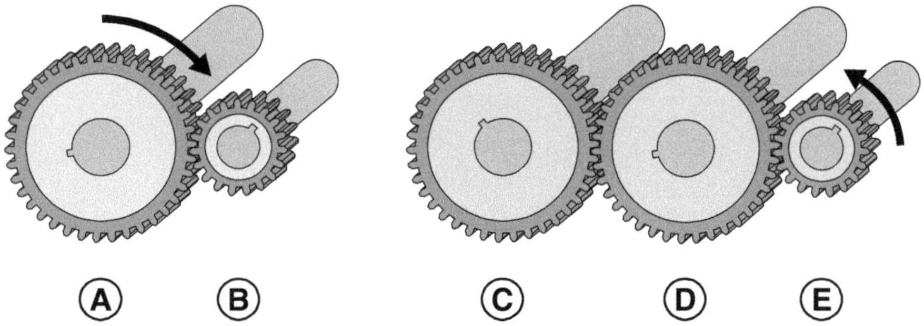

Activity: Lubrication

Oil from a gearbox with abrasive wear damage is analyzed.

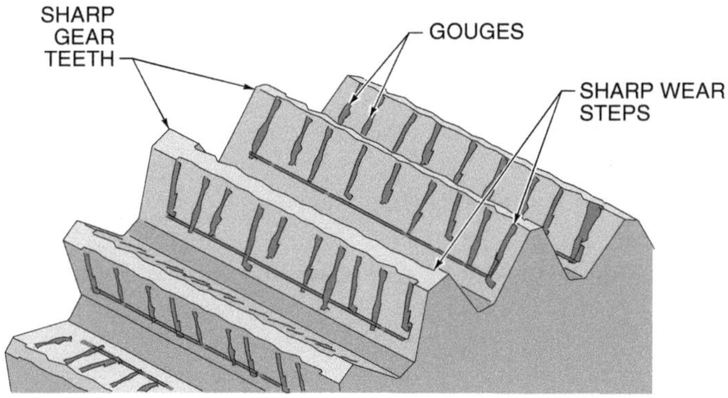

1. What type of contaminants would be found?

Corrosion is produced by water in the oil. The water causes rust and creates acid that removes metal from gear teeth.

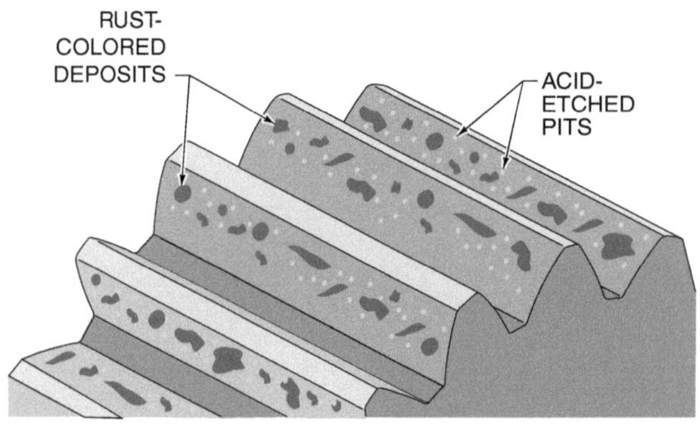

2. Which part of an oil analysis report would indicate that a gearbox has this type of contamination?

A new gearbox is installed, lubricated, and started for the first time.

3. What happens to the microscopic peaks and valleys left by the machining process during the break-in period?

4. What damage would occur if the microscopic layer of lubrication were disrupted and metal-to-metal contact were to occur?

5. Why might it be better to have just a few types of general-duty lubricants rather than a large number of specific-application lubricants?

6. Why should caps be kept on grease fittings?

Activity: Bearings

A bearing is installed with a pressing tool that applies force evenly around the bearing.

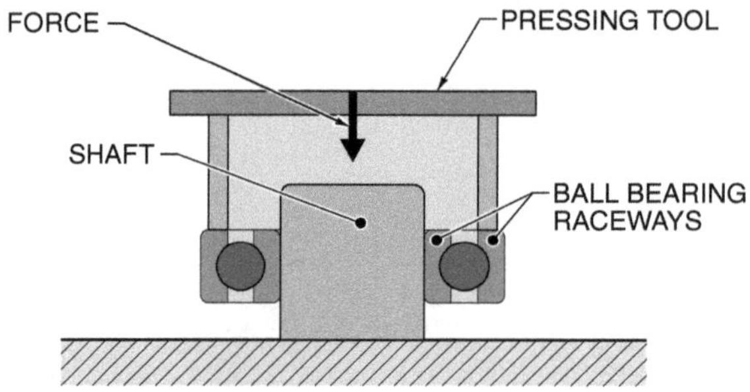

1. What problem will develop from this situation?

A bearing is installed correctly, but it continues to receive force from the pressing tool after it is seated in the housing.

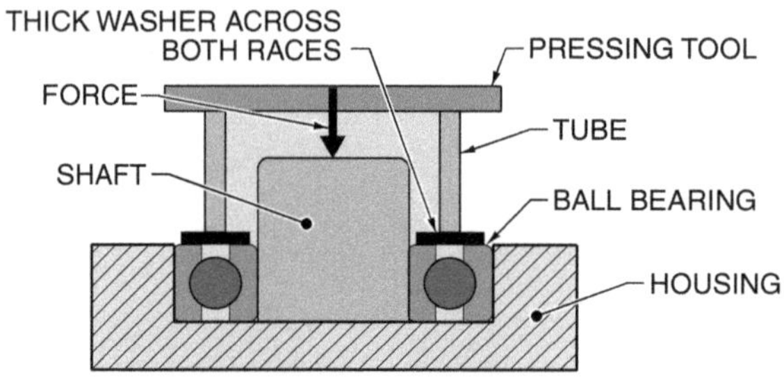

2. What problem will develop from this situation?

3. Why does overlubrication damage a bearing?

4. Why does most bearing wear occur at equipment startup?

5. Why is it important to try to remove a bearing without damaging it?

Activity: Mechanical Drive Systems

Answer the following questions about mechanical drive systems.

1. Why should sprockets be inspected when removing or inspecting a chain?

2. Why must all belts in a multiple-belt drive be changed when one belt is changed?

3. What is the purpose of analyzing the cause of a bearing, gear, chain, or belt drive failure?

4. How might the operation of a cam and cam follower be affected by misalignment?

5. Why are many variable-speed belt drives being replaced by electronic speed controls?

Activity: Belt Tensioning

Calculate the appropriate belt deflection for proper tension based on the following belt drive arrangements. There should be 1⁄64″ (0.0156″) of deflection from a straightedge for each inch of span length.

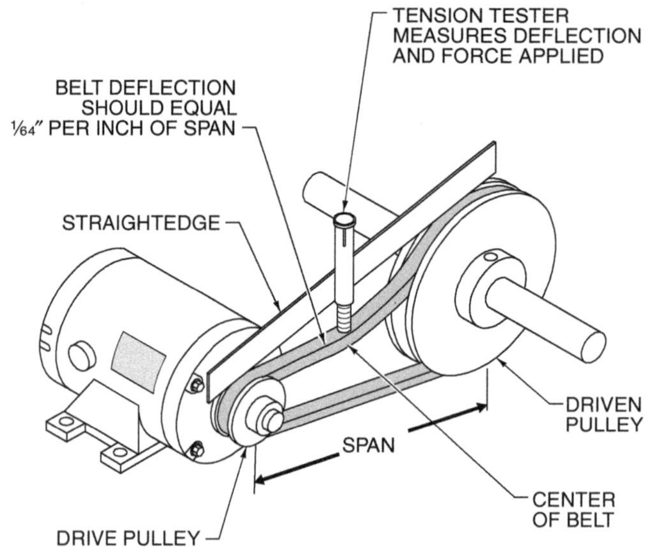

$h = L \times 1/64''$

where

h = deflection (in in.)

L = center-to-center span length between the pulleys (in in.)

1. The center-to-center distance of a belt drive is 12.00″.

2. The center-to-center distance of a belt drive is 48.50″.

3. The center-to-center distance of a belt drive is 21.75″.

4. The center-to-center distance of a belt drive is 33.25″.

5. Why must belt tension be checked again one or two days after installing new belts?

Activity: Alignment

A new motor is installed to operate a pump and must be checked for proper alignment.

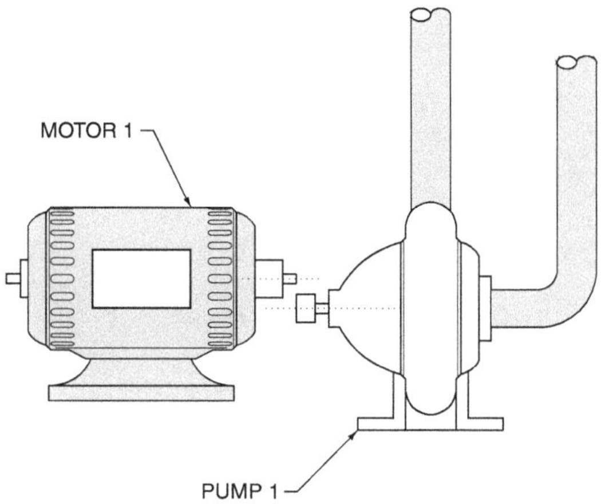

1. Indicate with an arrow or arrows how the motor must be moved to correct the alignment problem.

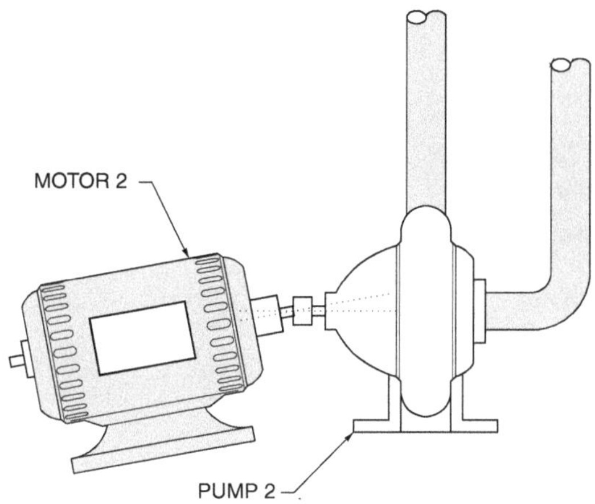

2. Indicate with an arrow or arrows how the motor must be moved to correct the alignment problem.

3. Could shaft misalignment add to the current draw of a motor driving a load?

4. What might an oil stain near the shaft coming out of a gearbox indicate?

5. Why might string be used to help align belt pulleys instead of a straightedge?

Activity: Gear Ratios

Answer the following questions based on the gear drive and the gear ratio formulas.

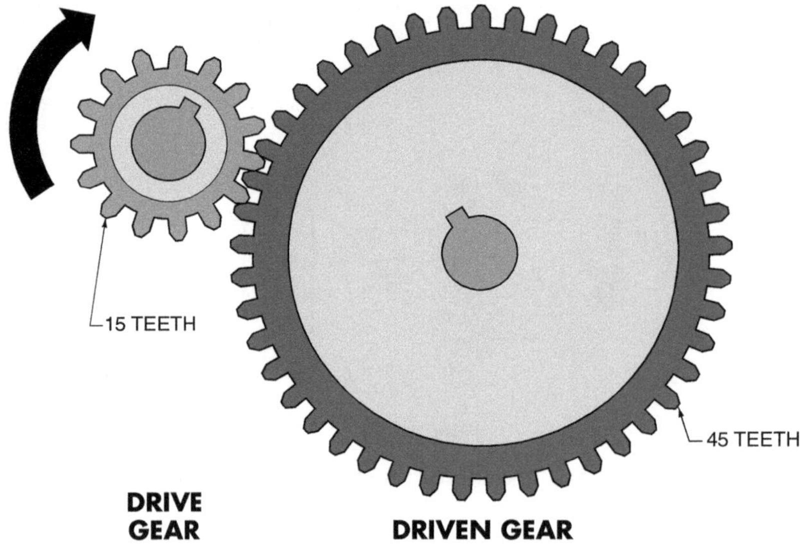

$$M = \frac{n_{drive}}{n_{driven}} = \frac{d_{drive}}{d_{driven}} = \frac{\omega_{driven}}{\omega_{drive}} = \frac{\tau_{drive}}{\tau_{driven}}$$

where

M = gear ratio
n_{drive} = number of teeth on drive gear
n_{driven} = number of teeth on driven gear
d_{drive} = diameter of drive gear
d_{driven} = diameter of driven gear
ω_{driven} = speed of driven gear
ω_{drive} = speed of drive gear
τ_{drive} = torque of drive gear
τ_{driven} = torque of driven gear

1. Which direction would the driven gear rotate (clockwise or counterclockwise)?

2. What is the gear ratio of this gear drive?

3. Will the large gear rotate faster or slower than the small gear?

4. If the torque on the drive gear is 3.5 lb-ft, what is the resulting torque on the driven gear?

Activity: Speed Ratios

Answer the following questions based on the belt drive and the speed ratio formulas.

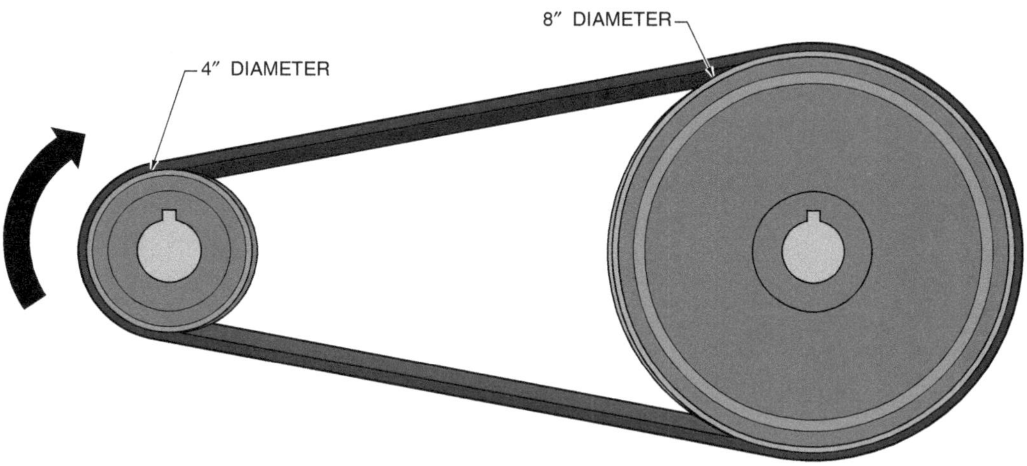

$$M = \frac{d_{drive}}{d_{driven}} = \frac{\omega_{driven}}{\omega_{drive}} = \frac{\tau_{drive}}{\tau_{driven}}$$

where
M = gear ratio
d_{drive} = diameter of drive gear
d_{driven} = diameter of driven gear
ω_{driven} = speed of driven gear
ω_{drive} = speed of drive gear
τ_{drive} = torque of drive gear
τ_{driven} = torque of driven gear

1. Is this belt drive a synchronous or nonsynchronous drive?

2. Which direction would the driven pulley rotate (clockwise or counterclockwise)?

3. What is the speed ratio of this belt drive?

4. Which pulley will rotate faster?

5. Which pulley will have greater torque?

6. If the driven pulley becomes jammed, what will likely happen to the belt?

ELECTRICAL PARAMETERS

SECTION 6.1

REVIEW QUESTIONS

Name ______________________________ **Date** ______________

True-False

T F **1.** An atomic nucleus consists of balanced numbers of protons and neutrons and, therefore, has no electrical charge.

T F **2.** The tendency for electrons to jump from one atom to another makes electricity possible.

T F **3.** An ampere is a unit of measure related to a certain number of electrons passing a given point per second.

T F **4.** Ohm's law states that current in a circuit is proportional to the voltage and the resistance.

T F **5.** Electrical power is measured in watts (W).

Multiple Choice

______________ **1.** The power formula is the relationship between ___ in a circuit.
- A. power, polarity, and conductivity
- B. power, resistance, and voltage
- C. power, resistance, and current
- D. power, voltage, and current

______________ **2.** A conductor is a material that has little ___.
- A. resistance
- B. electrical charge
- C. insulation
- D. polarity

______________ **3.** A(n) ___ is a subatomic particle that has a positive electrical charge of one unit.
- A. proton
- B. neutron
- C. electron
- D. atom

_______________ **4.** Current is measured in ___.
A. volts
B. amperes
C. ohms
D. watts

_______________ **5.** A(n) ___ is a subatomic particle that has a negative electrical charge of one unit.
A. proton
B. neutron
C. electron
D. atom

Completion

_______________ **1.** ___ current is current that reverses its direction of flow at regular intervals.

_______________ **2.** ___ is the amount of electrical pressure (electromotive force) that causes electrons to move in a circuit.

_______________ **3.** A(n) ___ electron is an electron in the outermost shell of an atom.

_______________ **4.** ___ is the opposition to the flow of electrons.

_______________ **5.** A(n) ___ is a material that has little resistance and permits electrons to move through it easily.

_______________ **6.** A(n) ___ is the smallest building block of matter that cannot be divided into a smaller unit without changing its basic character.

_______________ **7.** ___ is the flow of electrons through a conductor.

_______________ **8.** ___ is the positive or negative state of an object.

REVIEW QUESTIONS

Name ______________________________ Date ______________

True-False

T F 1. No electrical work is done in an open circuit.

T F 2. Voltage is the same throughout a series circuit.

T F 3. Ampacity is the amount of current that a conductor can safely carry.

T F 4. Larger AWG numbers indicate larger diameter conductors.

T F 5. Most industrial electrical circuits are strictly parallel circuits.

Multiple Choice

______________ **1.** A ___ is a device that uses electricity, such as a lamp, heating element, or motor, to produce a useful output.
- A. conductor
- B. circuit
- C. load
- D. breaker

______________ **2.** ___ is the reduction of circuit voltage caused by the resistance of circuit conductors and loads.
- A. Voltage drop
- B. Short circuiting
- C. Insulation
- D. Grounding

______________ **3.** Circuit ___ is the set of operational characteristics of a circuit as determined by the arrangement of switch devices.
- A. display
- B. functionality
- C. distribution
- D. logic

_______________ **4.** A grounding conductor has ___ insulation or is bare and is connected to a metal grounding electrode.
A. white
B. green
C. grey
D. black

_______________ **5.** Two or more switches connected in parallel form what is called ___ logic.
A. AND
B. OR
C. SERIES
D. SAFETY

Completion

_______________ **1.** A(n) ___ circuit is a circuit having an incomplete path, which prevents current flow.

_______________ **2.** A(n) ___ circuit is an undesirable, low-resistance path for current to bypass a load.

_______________ **3.** A(n) ___ circuit is a circuit that has two or more components connected with only one path for current flow.

_______________ **4.** A(n) ___ circuit is a circuit having a complete path for current flow.

_______________ **5.** A(n) ___ circuit is a low-resistance connection between electrical equipment and the earth.

POWER DISTRIBUTION

SECTION 6.3

REVIEW QUESTIONS

Name ______________________________ **Date** ____________________

True-False

T F **1.** A generator is an electrical component that converts AC to DC by allowing voltage and current to flow in only one direction.

T F **2.** Three-phase circuits do not always require a neutral wire.

T F **3.** All points in an AC circuit have polarity.

T F **4.** In three-phase AC voltage, the coils are out-of-phase with each other.

T F **5.** Larger emergency generators are usually powered by diesel engines.

Multiple Choice

______________ **1.** A ___ is an electric device that converts mechanical energy into electrical energy by means of electromagnetic induction.
- A. generator
- B. panelboard
- C. rectifier
- D. load

______________ **2.** Frequency is measured in cycles per second, or ___.
- A. volts
- B. waves
- C. watts
- D. hertz

______________ **3.** Three-phase AC voltage is a combination of three alternating voltage waveforms, each displaced ___ electrical degrees apart.
- A. 60
- B. 90
- C. 120
- D. 180

_______________ **4.** A ___ is a piece of equipment that breaks down incoming electrical power into smaller units for distribution throughout a building.

A. disconnect

B. switchboard

C. lighting panel

D. busway

Completion

_______________ **1.** A(n) ___ is an electrical device that steps up or steps down the voltage of AC power.

_______________ **2.** From the main switchboard, ___ are used to distribute electricity for feeder circuits.

_______________ **3.** Power ___ is the process of delivering electrical power to where it is needed.

_______________ **4.** The United States distributes AC voltage at a frequency of ___ Hz.

_______________ **5.** High-voltage 1ϕ power (240 V) is delivered with ___ energized conductors.

ELECTRICAL TEST INSTRUMENTS

SECTION **6.4**

REVIEW QUESTIONS

Name ______________________________ **Date** ________________

True-False

T	F	**1.** Voltage measurements are used to determine the amount of circuit loading or the condition of an electrical component (load).
T	F	**2.** A clamp-on probe does not need to touch a conductor to measure current.
T	F	**3.** Always disconnect the potentially energized (red) test lead first.
T	F	**4.** Location is not factor that affects CAT ratings.
T	F	**5.** Voltage indicators do not indicate the type or level of voltage, only that voltage is present.

Multiple Choice

______________ **1.** A ___ is a test instrument that indicates the presence of voltage when the test tip is near an energized conductor or metal part.

A. circuit analyzer
B. voltage indicator
C. digital multimeter
D. megohmmeter

______________ **2.** A ___ is used to measure the very high resistances of insulating components.

A. digital multimeter
B. continuity tester
C. megohmmeter
D. circuit analyzer

______________ **3.** A(n) ___ value is the result of a sequence of calculations applied to a changing value that provides an overall, or effective, representation of the waveform.

A. root-mean-square
B. peak
C. MIN/MAX
D. relative

______________ 4. The ___ function establishes a baseline value and displays the current measurement as a positive or negative difference from that baseline.

A. MIN/MAX
B. HOLD
C. PEAK
D. RELATIVE

______________ 5. Resistance is displayed on a meter as ___.

A. ohms
B. volts
C. watts
D. amperes

Completion

______________ 1. ___ is the recording of measurements at specified intervals over time.

______________ 2. A(n) ___ instrument is an electrical measurement tool used to test the condition or operation of an equipment component or system.

______________ 3. A(n) ___ is a test instrument capable of measuring two or more electrical quantities and displaying the exact values on a digital display.

______________ 4. ___ is the presence of a completed path for current flow.

______________ 5. A(n) ___ value is the maximum instantaneous value of either the positive or negative alternation of a sine wave.

POWER QUALITY

SECTION **6.5**

REVIEW QUESTIONS

Name ______________________________ **Date** ______________________

True-False

T F **1.** Small voltage fluctuations typically cause significant circuit and load problems.

T F **2.** The fundamental line frequency in the United States is 60 Hz.

T F **3.** Lightning is the most common source of transient voltages on utility power distribution systems.

T F **4.** When voltage fluctuations are found to be more than ±8%, a surge suppressor must be added to the system.

T F **5.** Typically, for every 1% of voltage unbalance, current unbalance is 4% to 8%.

Multiple Choice

______________ **1.** A ___ voltage is a temporary, undesirable voltage spike that ranges from a few volts to several thousand volts and lasts from a few microseconds up to a few milliseconds.

A. transverse
B. transient
C. transparent
D. translucent

______________ **2.** A voltage sag is a drop in voltage of more than ___% (but not to 0 V) below the normal rated line voltage that lasts from 0.5 cycles up to 1 min.

A. 2
B. 5
C. 10
D. 25

______________ **3.** A voltage ___ is an increase in voltage of more than 10% above the normal rated line voltage that lasts from 0.5 cycles up to 1 min.

A. wave
B. surge
C. transient
D. swell

_______________ **4.** A ___ power interruption is a decrease in voltage to 0 V that lasts from 3 sec to 1 min.

A. momentary
B. sustained
C. temporary
D. dropped

_______________ **5.** ___ is the complete loss of one phase on a 3ϕ power supply.

A. Voltage unbalance
B. Single phasing
C. Phase unbalance
D. Improper phase sequence

Completion

_______________ **1.** A(n) ___ is a drop in voltage by a power company to conserve power during times of peak usage or excessive loading of the power distribution system.

_______________ **2.** A(n) ___ is a device that limits the intensity of voltage transients that occur in a power distribution system.

_______________ **3.** ___ is the changing of the sequence of any two phases in a 3ϕ system or circuit.

_______________ **4.** ___ is the addition of voltage and/or current in a power line that is a multiple of the fundamental line frequency.

_______________ **5.** ___ power interruptions (outages) are commonly the result of storms or circuit breakers tripping due to damaged equipment.

ELECTRICAL DEVICES

SECTION 6.6

REVIEW QUESTIONS

Name ______________________________ **Date** ____________________

True-False

T F **1.** A bimetallic strip is made of two dissimilar metals that expand at the same rate when heated.

T F **2.** High-intensity discharge (HID) lamps require several minutes to begin emitting light and most produce a humming sound from their ballasts.

T F **3.** An induction motor is a motor that rotates due to the interaction between the magnetic fields of the stator and rotor.

T F **4.** In a step-up transformer, the secondary coil has more turns of wire than the primary coil.

T F **5.** A solenoid is a relay designed for high-current loads.

Multiple Choice

______________ **1.** A ___ is an OCPD with a metal strip that quickly melts and opens a circuit when the current flowing through the strip is greater than the rating of the fuse.
- A. CB
- B. fuse
- C. switch
- D. lamp

______________ **2.** ___ occurs when contacts do not close, do not open, or offer resistance to current flow.
- A. Power interruption
- B. Voltage drop
- C. Time delay
- D. Contact failure

______________ **3.** A(n) ___ switch is a switch that is operated by the movement of an object.
- A. manual
- B. mechanical
- C. automatic
- D. semi-automatic

______________ **4.** A(n) ___ lamp is an electric lamp that produces light as electricity passes through a gas.
A. fluorescent
B. incandescent
C. light-emitting diode (LED)
D. electroluminescent

______________ **5.** A(n) ___ is the moving part of a coil-operated device.
A. drive
B. rotor
C. armature
D. relay

Completion

______________ **1.** A(n) ___ switch is a switch that is operated by a person.

______________ **2.** A(n) ___ is a device that changes DC voltage into AC voltage of any frequency.

______________ **3.** A(n) ___ is an electrical switch that is actuated by a separate circuit.

______________ **4.** A(n) ___ is any device that automatically opens a circuit to prevent damage from an overcurrent condition.

______________ **5.** A(n) ___ is a conducting part of a switch that operates with another conducting part to complete (close) or break (open) a circuit.

SECTION **6.7**

REVIEW QUESTIONS

Name ______________________________ **Date** ____________________

True-False

T F **1.** Line diagrams are commonly used for power circuits, and wiring diagrams are commonly used for low-voltage control circuits.

T F **2.** Solid black dots at line intersections mean that the wires are connected electrically.

T F **3.** Frequently, solid lines are drawn around certain sections of a wiring diagram to identify that portion as a group.

Multiple Choice

________________ **1.** In a diagram, wires crossing without a dot, or with one wire drawn as "jumping" over the other, indicates that there is ___.

A. a terminal connected to both wires
B. no electrical connection
C. a splice connecting the wires together
D. a mistake in the wiring

________________ **2.** By convention, terminal 1 is always the ___ source conductor.

A. ungrounded
B. grounded
C. neutral
D. cold

________________ **3.** The wiring diagram of a standard start-stop station shows the motor starter and its wiring with ___ lines to identify it as part of the power circuit.

A. ladder
B. red
C. dashed
D. thick

Completion

________________ **1.** A(n) ___ diagram is a diagram that uses lines and graphic symbols to show the logic and operation of an electrical circuit.

____________________ **2.** A standard ___ station used to control a single small- to medium-sized motor is one of the most common industrial control circuits.

____________________ **3.** A(n) ___ diagram is a diagram that shows the electrical connections of a circuit.

ELECTRICAL SYSTEM MAINTENANCE

SECTION **6.8**

REVIEW QUESTIONS

Name ______________________ **Date** ______________

True-False

T F **1.** Overheating is the number one cause of electrical equipment failure.

T F **2.** Motor current draw should be checked annually.

T F **3.** When using a megohmmeter, the motor must be connected to its circuit for testing.

T F **4.** The combination of dirty walls and dirty lamp fixtures can reduce light output by up to 50%.

Multiple Choice

______________ **1.** ___ cause high resistance, leading to dangerous overheating and open circuits.
A. Overcurrents
B. Contact failures
C. Vibrations
D. Loose connections

______________ **2.** To prevent insulation from overheating, ___ should be used to carefully remove any dirt or foreign matter from inside the motor.
A. low-pressure compressed air
B. a dry cloth
C. clean, warm water
D. a dry-bristle brush

______________ **3.** During an insulation spot test, a ___ is connected to each motor winding lead (in turn) and to ground, and the resistance is measured after 60 sec.
A. continuity tester
B. megohmmeter
C. phase tester
D. ground resistance tester

______________ **4.** Group ___ is the replacement of all lamps in a given area when they reach 60% to 80% of their rated average life.
A. testing
B. relamping
C. monitoring
D. fixturing

Completion

______________________ **1.** In an insulation step-voltage test, the resistance of good insulation remains approximately the same at different ___ levels.

______________________ **2.** Humming or chattering is an indication of mechanical problems interfering with the closing of ___.

______________________ **3.** A(n) ___ is a test performed to determine insulation resistance.

______________________ **4.** A(n) ___ is an insulation resistance test that identifies wet or contaminated insulation.

ELECTRICAL SYSTEM TROUBLESHOOTING

SECTION 6.9

REVIEW QUESTIONS

Name ________________________________ Date ________________

True-False

T F 1. A bad coil indicates OL or near zero resistance.

T F 2. A motor is normally loaded if the motor current is between 95% and 105% of the rated current.

T F 3. Intermittent problems are easier to diagnose than constant problems.

T F 4. Overheating is the most common cause of motor failure.

T F 5. All lamp ballasts produce some noise.

T F 6. Overheating in a coil can be caused by voltage that is 2% above or below its rated voltage.

Multiple Choice

______________ 1. The ___ testing method is a testing method in which one test instrument probe is connected to a point on a circuit (typically the grounded conductor) and the other probe is moved along the circuit to test sections sequentially.

A. spot
B. tie-down
C. source voltage
D. sliding

______________ 2. A good circuit breaker indicates ___ when the contacts are closed and OL when open.

A. nearly 0 Ω
B. 10 Ω to 20 Ω
C. about 50 Ω
D. OL

______________ 3. In general, a branch circuit should have a voltage drop of no more than ___% of the source voltage.

A. 3
B. 8
C. 12
D. 15

______________ **4.** ___ is the discharge of an electric current across a gap.
A. Stabbing
B. Pitting
C. Arcing
D. Tracing

______________ **5.** ___ is one method for raising ballast temperature.
A. Adding more lamps
B. Increasing the voltage
C. Replacing the lamps
D. Enclosing the complete fixture

______________ **6.** Voltage should be within ±___% of the required motor voltage.
A. 2
B. 10
C. 35
D. 50

______________ **7.** ___ is the percentage of time a load or circuit is on.
A. Loading
B. Average work rating
C. Duty cycle
D. Operational percentage

Completion

______________ **1.** A(n) ___ short is a short circuit that opens an overcurrent protective device (OCPD) as soon as a circuit is energized.

______________ **2.** The location of a short is indicated by signs of ___, such as burn marks or discolored insulation.

______________ **3.** A defective heating element with an open indicates ___ when connected to a DMM set to measure resistance.

______________ **4.** High-resistance ___ are often caused by damaged, but not completely broken, insulation or oil-soaked debris in a junction box where wires are connected.

______________ **5.** A normally open (NO) contact in good condition indicates ___ voltage when open (not actuated).

______________ **6.** A(n) ___ is a device that is pressed, pulled, or rotated by the individual operating a circuit.

Chapter 6 ACTIVITIES

Name ______________________________ **Date** ____________________

Ohm's Law

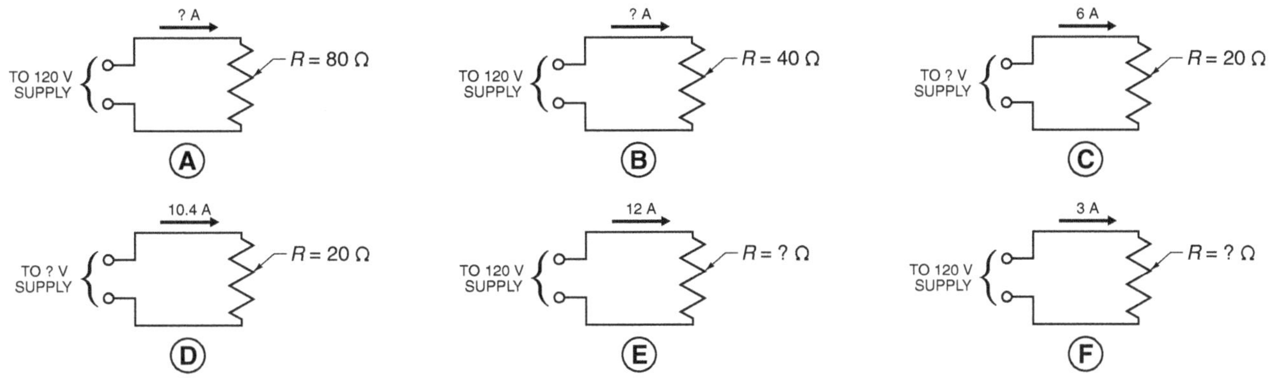

______________ **1.** The current at A is ___ A.

______________ **2.** The current at B is ___ A.

______________ **3.** The voltage at C is ___ V.

______________ **4.** The voltage at D is ___ V.

______________ **5.** The resistance at E is ___ Ω.

______________ **6.** The resistance at F is ___ Ω.

Copper Atom

______________ **1.** Electrons

______________ **2.** Shells

______________ **3.** Valence electron

______________ **4.** Nucleus

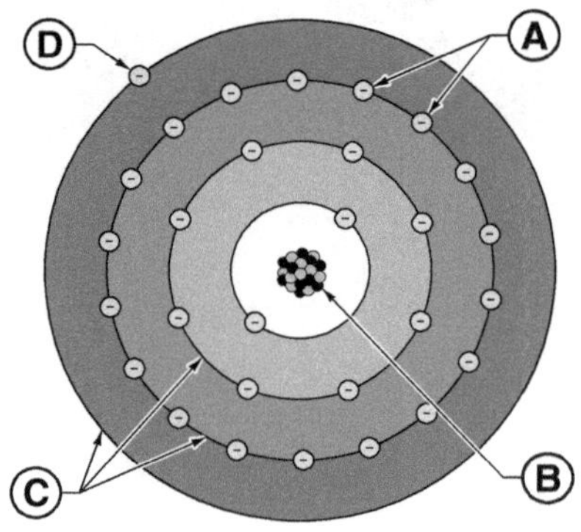

Three-Phase Motors

______________ **1.** Rotor

______________ **2.** Bearing

______________ **3.** Coils

______________ **4.** Shaft

______________ **5.** Fan

______________ **6.** Endbell

______________ **7.** Stator

______________ **8.** Motor mount

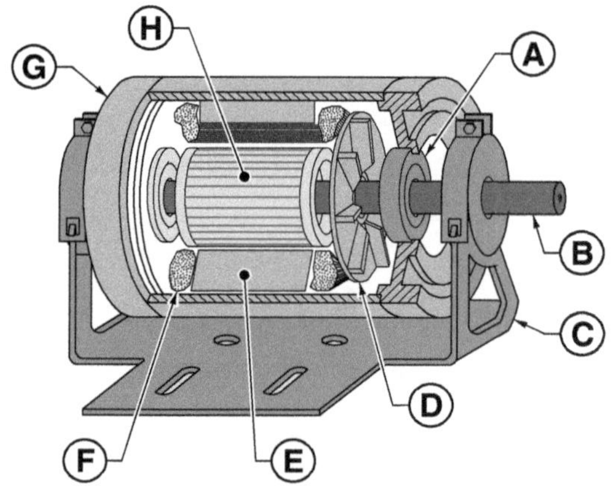

Power Formula

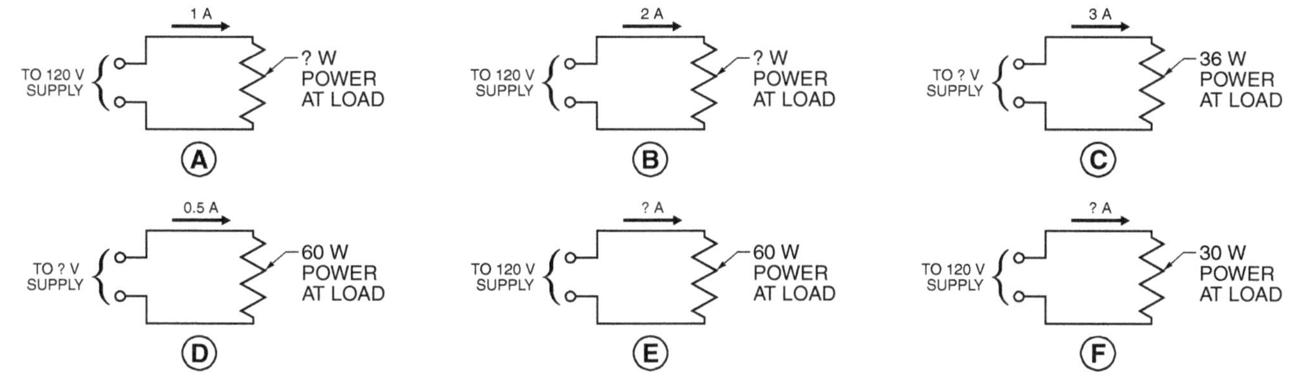

________________ **1.** The power at A is ___ W.

________________ **2.** The power at B is ___ W.

________________ **3.** The voltage at C is ___ V.

________________ **4.** The voltage at D is ___ V.

________________ **5.** The current at E is ___ A.

________________ **6.** The current at F is ___ A.

Magnetic Motor Starters

________________ **1.** Motor circuit contacts

________________ **2.** Armature

________________ **3.** Coil terminals

________________ **4.** Overload contact

________________ **5.** Heater unit

________________ **6.** Coil

________________ **7.** Auxiliary contacts

________________ **8.** Overload reset

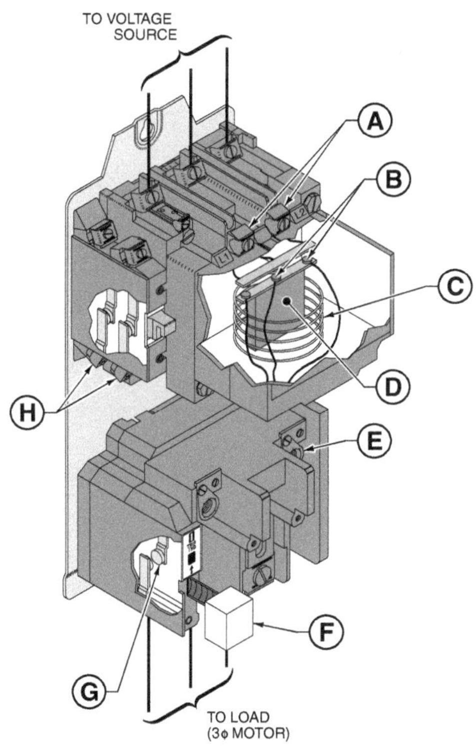

Circuit Measurements

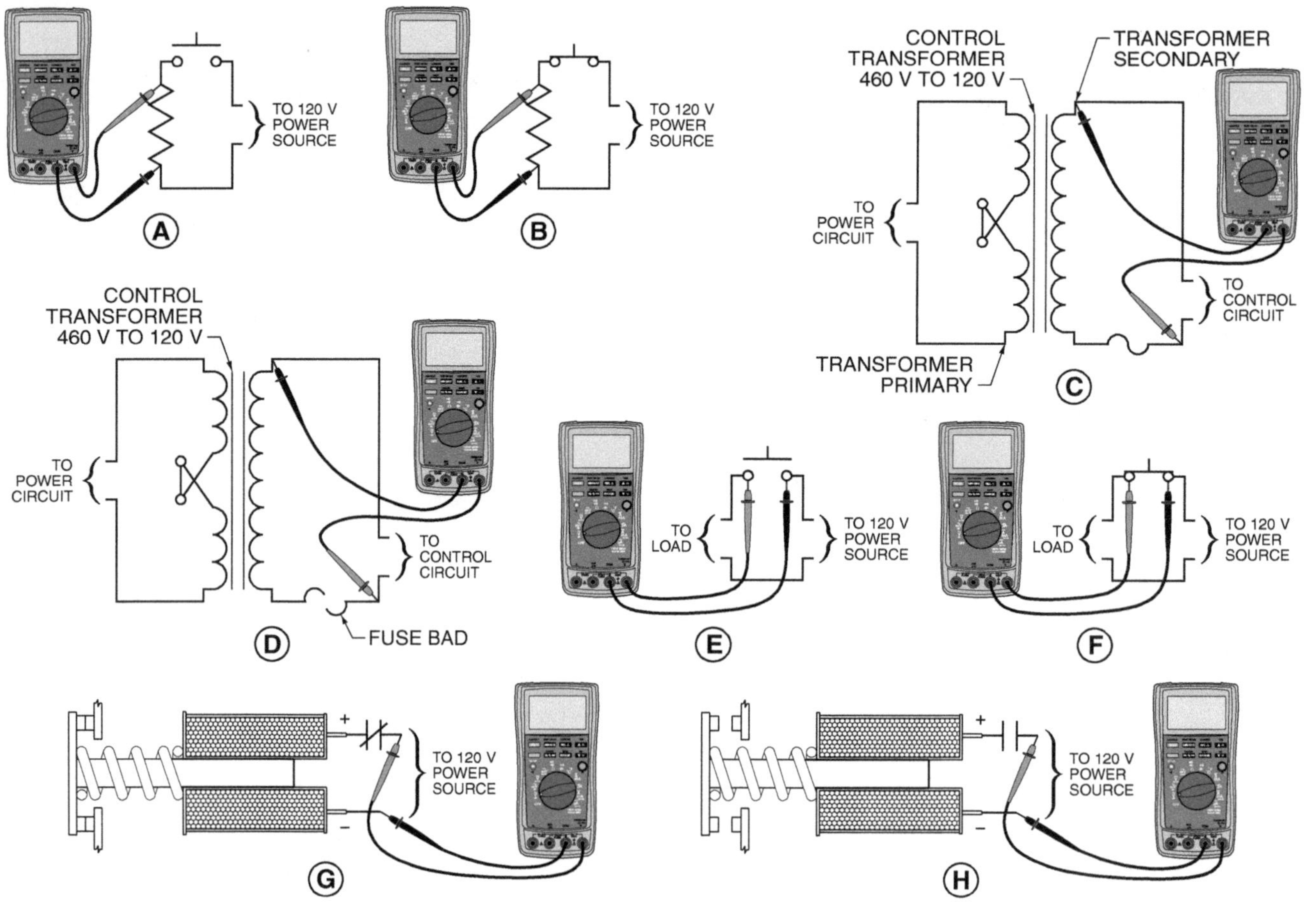

________________ **1.** The DMM at A reads ___ V.

________________ **2.** The DMM at B reads ___ V.

________________ **3.** The DMM at C reads ___ V.

________________ **4.** The DMM at D reads ___ V.

________________ **5.** The DMM at E reads ___ V.

________________ **6.** The DMM at F reads ___ V.

________________ **7.** The DMM at G reads ___ V.

________________ **8.** The DMM at H reads ___ V.

Activity: Reading Line Diagrams

A line diagram shows the operation of a control circuit. The circuit includes start and stop pushbuttons, motor coils, contacts, and overload protection. The M1, M2, and M3 coils operate magnetic motor starters. When a coil is energized, it causes the contacts to close, energizing a motor. The motor is not included in this diagram, only the controls that operate the magnetic motor starters.

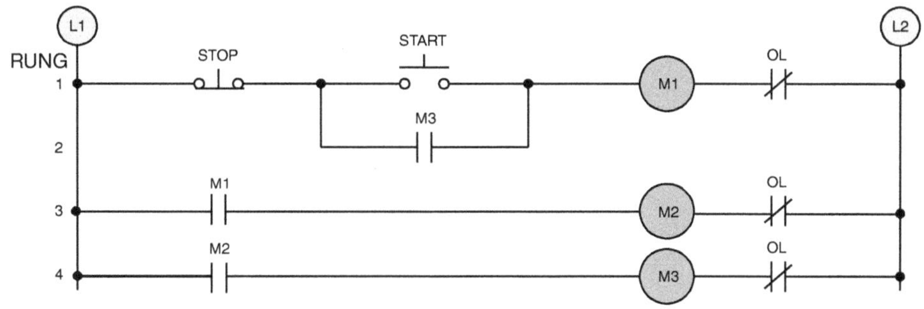

1. When the start button is pressed, which coil energizes first?

2. This causes the M1 contacts on which rung to close?

3. This causes which other coil to energize?

4. Does pressing the start button energize all three coils?

5. Will the coils remain energized when the start button is opened?

6. If the stop button is pressed, which coil(s) deenergize?

7. If the M2 overload contacts open, will the M1 coil deenergize?

A line diagram shows the operation of a control circuit. The circuit includes start and stop pushbuttons, a timer, motor coils, contacts, and overload protection.

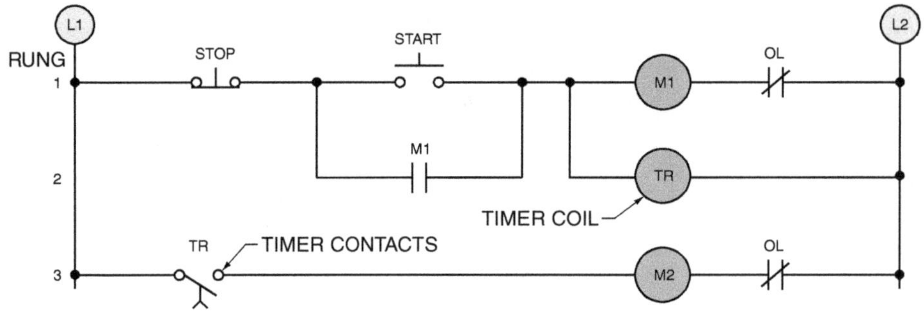

8. When the start button is pressed, which coil energizes first?

9. This causes the M1 contacts on which rung to close?

The closing of M1 contacts provides a holding circuit so that the M1 coil will remain energized when the start button is released. The motor being controlled by M1 energizes.

10. Which other coil energizes at the same time?

11. After the timer times out, which contacts close?

12. These contacts energize which coil?

13. If the stop button is pressed or the M1 overloads open, which coil(s) deenergize(s)?

14. If the M2 overload contacts open, will the M1 coil deenergize?

The M1 and M2 coils are inside motor starters. When the coils energize, the motors start.

15. If power to this circuit is lost, will the motors restart without pressing the start button again?

A line diagram shows the operation of a control circuit. The circuit includes a float switch, motor coil, and overload protection.

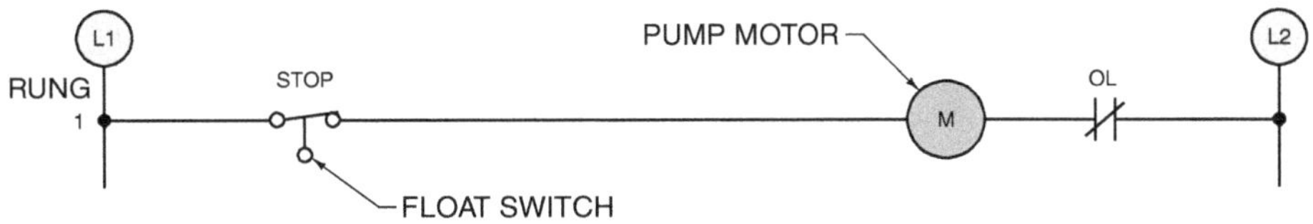

16. When the float switch is closed, which component is energized?

17. If the M overloads are open, will the motor restart if the float switch is closed?

18. If power to this circuit is lost, will the motor restart when power is restored, the float switch is closed, and the M overloads are closed?

19. What type of safety precautions should be taken when working on this type of automatic circuit?

A line diagram shows the operation of a control circuit. The circuit includes a float switch, a timer, timer contacts, a motor coil, and overload protection.

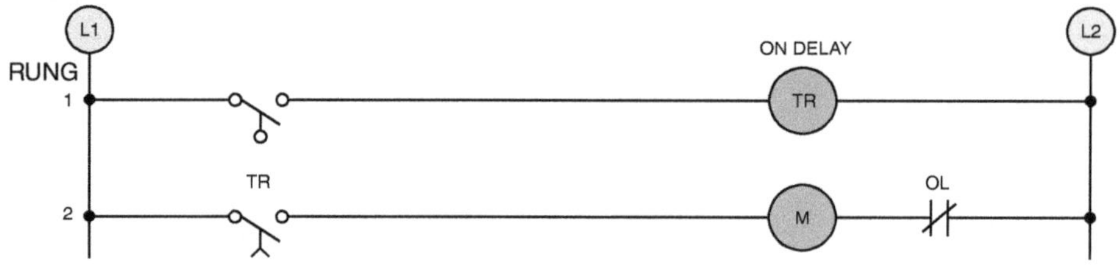

20. When the float switch closes, which coil energizes?

21. When the timer times out, which coil energizes?

22. When the float switch opens, which coil(s) deenergize(s)?

23. If the M overloads are open, will the motor restart if the float switch is closed?

24. If the M overloads are open, will the timer start if the float switch is closed?

25. If power to this circuit is lost, will the motor restart when power is restored, the float switch is closed, and the M overloads are closed?

Activity: Series Resistance in a Circuit

A line diagram shows several loads and switches connected to a power circuit.

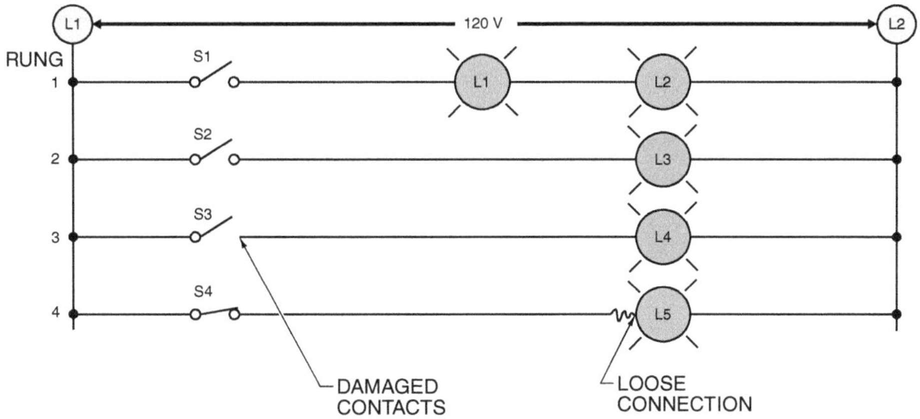

1. If both light bulbs in Rung 1 are the same wattage, what is the voltage drop across each light?

2. How would the brightnesses of the light bulbs in Rung 1 compare to each other?

3. How would their voltage drops and brightnesses compare if L1 were a higher wattage than L2?

4. If the switch contacts are clean, what is the voltage drop across S2 when it is closed?

5. What is the approximate voltage drop across L3 in Rung 2?

6. How do the problems in Rung 3 and Rung 4 affect the voltage available to L4 and L5?

7. What would happen to the brightness of light bulbs L4 and L5?

8. What voltage drop reading indicates damaged contacts?

9. What resistance readings indicate dirty contacts?

10. What maintenance activities will help prevent loose connections or dirty contacts?

Activity: Parallel Resistance in a Circuit

A line diagram shows several loads and switches connected to a power circuit.

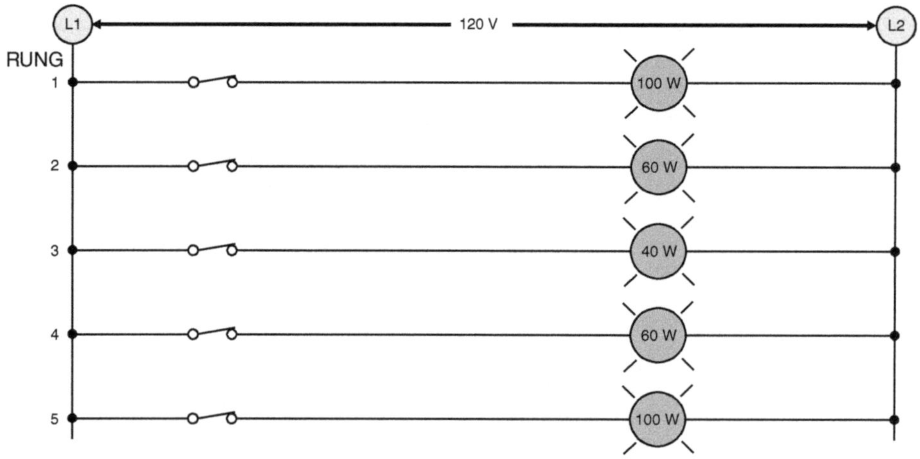

1. What happens to total circuit resistance if individual rungs are disconnected one at a time, starting with Rung 5?

2. What happens to total circuit resistance when loads are added to a parallel circuit?

3. What happens to total circuit current when loads are added to a parallel circuit?

4. What must happen to circuit current flow in order for a fuse to blow or a circuit breaker to trip?

5. If the circuit were supplied with 100 V instead of 120 V, what would happen to total circuit current?

6. Which rung(s) have the least current flow?

7. Which rung(s) have the highest current flow?

8. What difference must there be between Rung 2 and Rung 4 if current flow were lower in Rung 2 than in Rung 4?

9. What would happen to a load rated for 25 V if it were supplied with 120 V?

10. What would happen to a load rated for 208 V if it were supplied with 120 V?

11. What must be done when a fuse blows or a circuit breaker opens for no apparent reason?

Activity: Voltage Unbalance

Calculate the voltage unbalance in the following three-phase circuits.

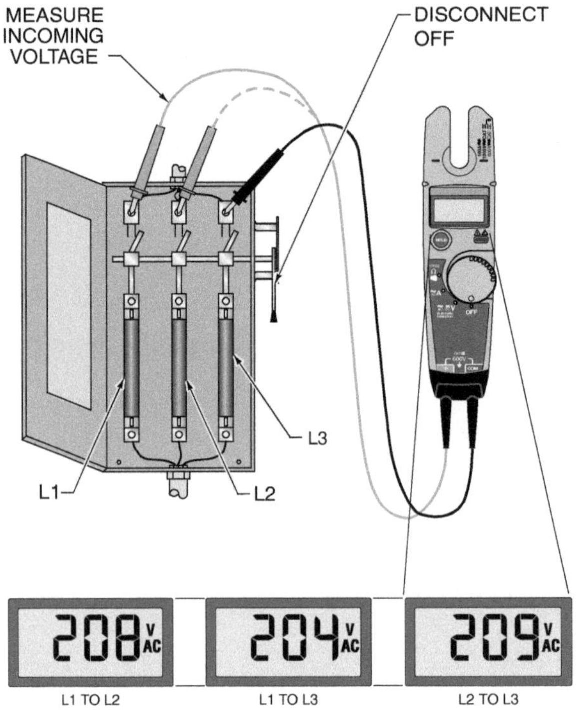

1. What is the average voltage?

2. What is the maximum deviation from the average?

3. What is the percentage of unbalance (the percentage of maximum deviation from the average)?

4. Is this circuit acceptably balanced?

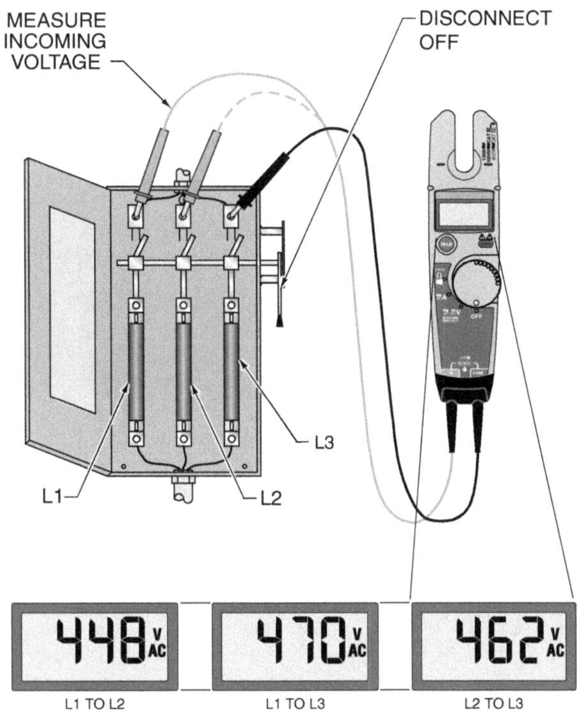

5. What is the average voltage?

6. What is the maximum deviation from the average?

7. What is the percentage of unbalance (the percentage of maximum deviation from the average)?

8. Is this circuit acceptably balanced?

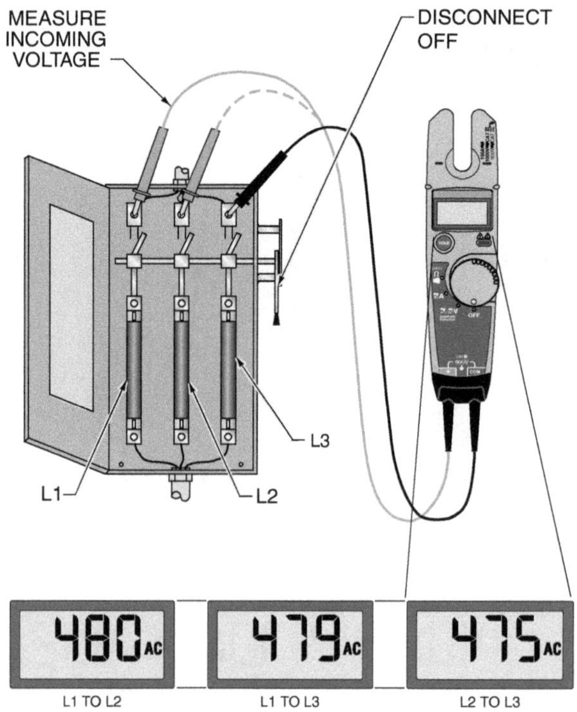

9. What is the average voltage?

10. What is the maximum deviation from the average?

11. What is the percentage of unbalance (the percentage of maximum deviation from the average)?

12. Is this circuit acceptably balanced?

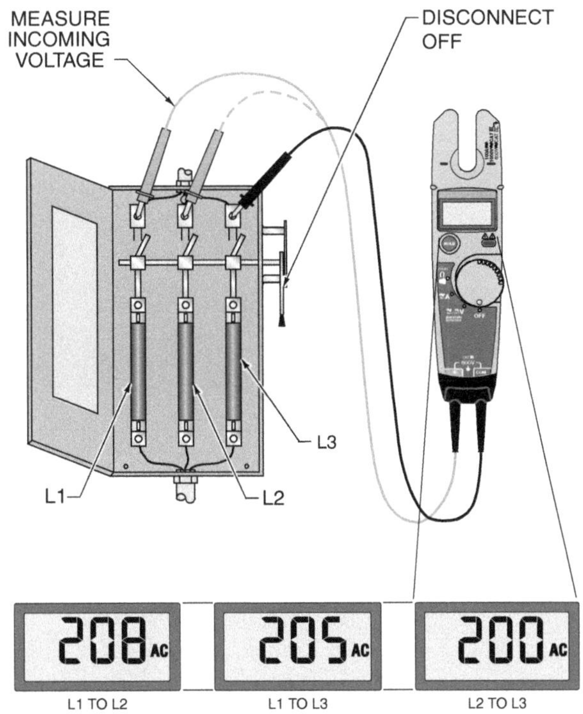

13. What is the average voltage?

14. What is the maximum deviation from the average?

15. What is the percentage of unbalance (the percentage of maximum deviation from the average)?

16. Is this circuit acceptably balanced?

Activity: Current Unbalance

Calculate the current unbalance in the following three-phase circuits.

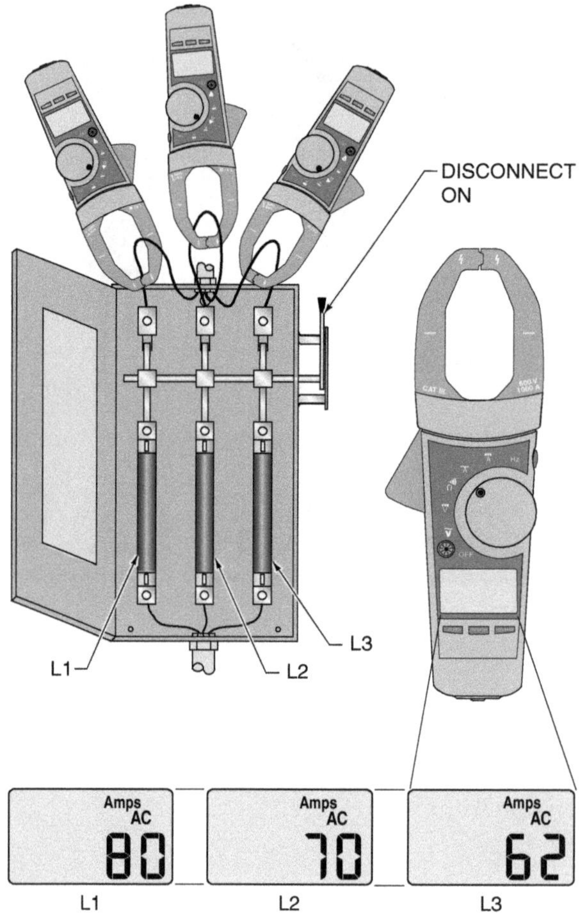

1. What is the average current?

2. What is the maximum deviation from the average?

3. What is the percentage of unbalance (the percentage of maximum deviation from the average)?

4. Is this circuit acceptably balanced?

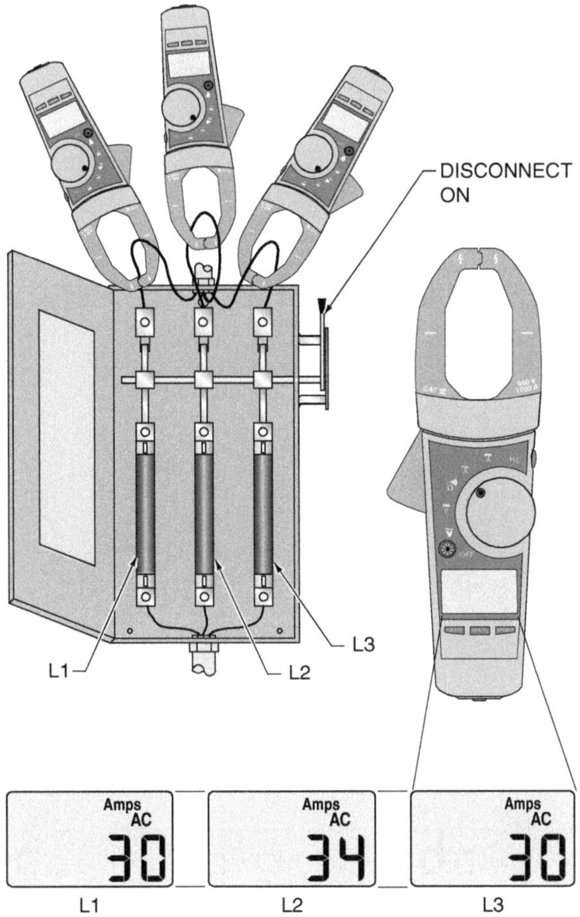

5. What is the average current?

6. What is the maximum deviation from the average?

7. What is the percentage of unbalance (the percentage of maximum deviation from the average)?

8. Is this circuit acceptably balanced?

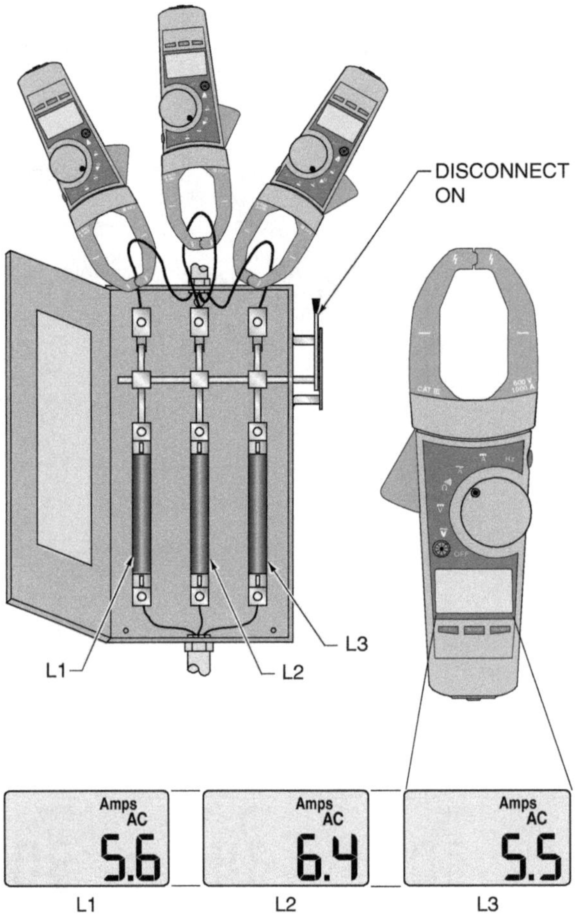

9. What is the average current?

10. What is the maximum deviation from the average?

11. What is the percentage of unbalance (the percentage of maximum deviation from the average)?

12. Is this circuit acceptably balanced?

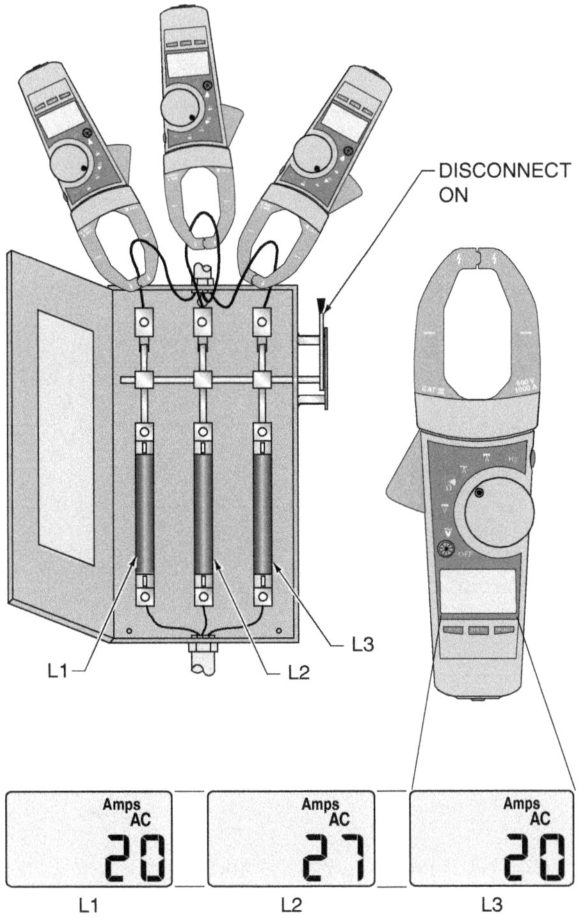

13. What is the average current?

14. What is the maximum deviation from the average?

15. What is the percentage of unbalance (the percentage of maximum deviation from the average)?

16. Is this circuit acceptably balanced?

Activity: Testing for Opens

A line diagram shows several loads and switches connected to a power circuit. Opens in circuits and loads are a common problem in electrical and electronic systems. Locating single and multiple opens in a circuit can be simplified by using sensitive, high-quality meters.

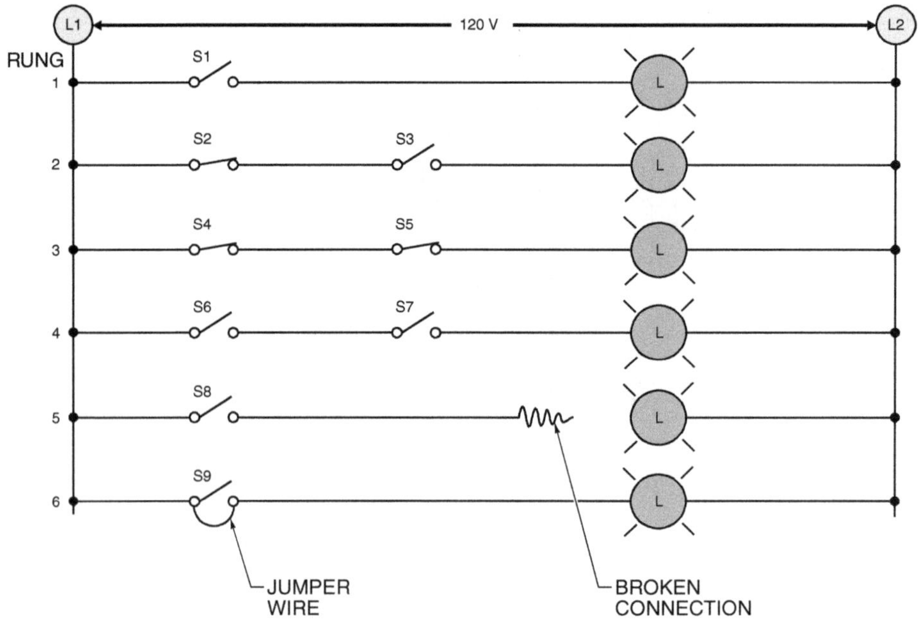

1. What is the voltage drop across S1 when it is closed and the voltage potential when it is open?

2. What is the voltage drop across S2 when it is closed and S3 is open?

3. What is the voltage drop across S4 when it is closed and S5 is closed?

4. What is the approximate voltage potential across S6 when both S6 and S7 are open?

5. What is the approximate voltage potential across S7 when both S6 and S7 are open?

6. What is the voltage potential across S8 when it is open and there is a broken connection to the load?

7. What would happen if a jumper wire were placed across S9?

8. What dangerous situation can this cause?

9. If S9 is not closing when the switch is operated, what would applying the jumper wire across S9 confirm?

10. How can a high-quality meter be used to detect multiple opens in series when troubleshooting circuits?

Activity: Testing for Opens in an Industrial Circuit

A holding tank is filled by a pump, which is controlled by float switches that turn the pump on and off. The switches are mounted in the tank at various levels. A low-level switch turns the pump on if the liquid level falls below it. A high-level switch turns the pump off if the liquid level rises above it. An overfill switch sounds an alarm if the pump does not stop and the liquid level reaches the overfill level. Sensitive voltmeters (or multimeters set to measure voltage) are used to test the control circuit and the switches for opens. The circuit operates on 120 V.

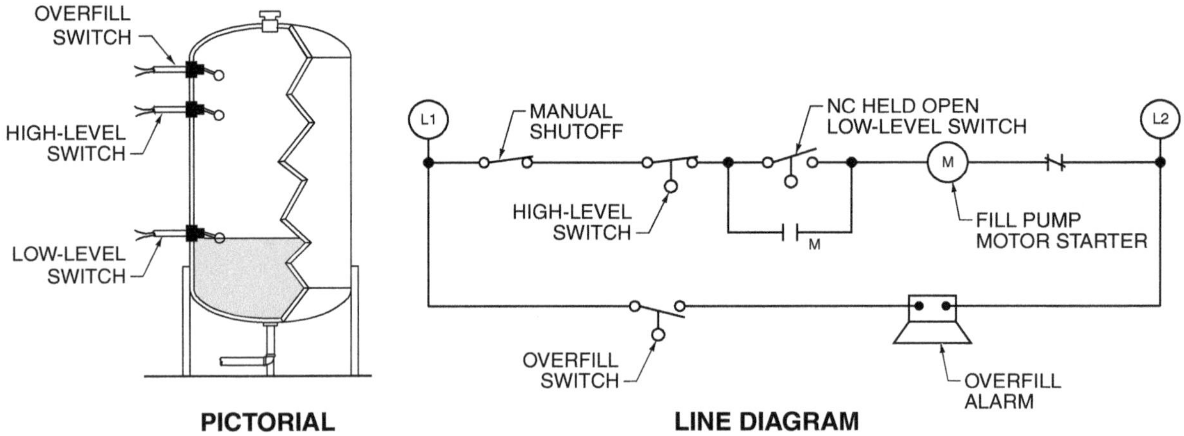

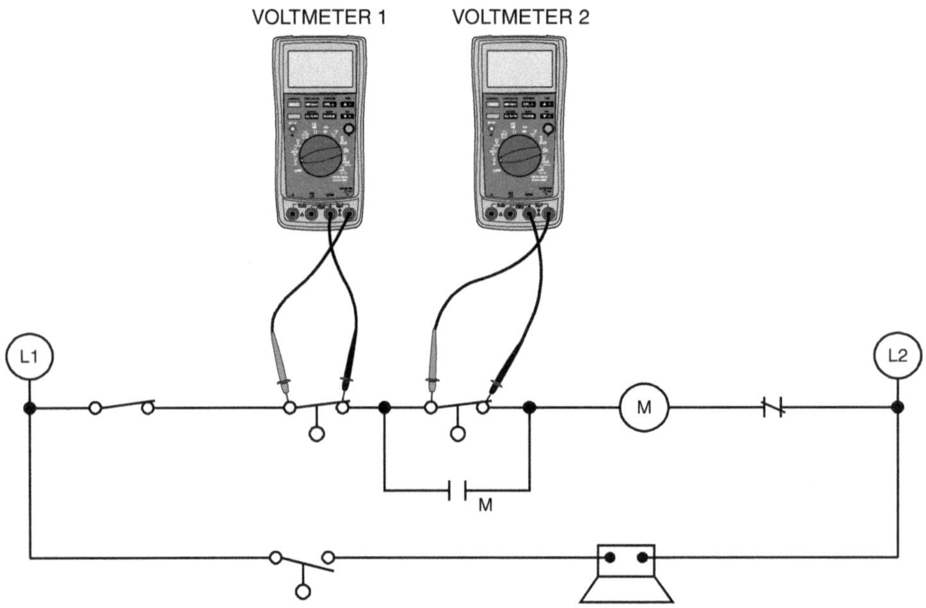

1. What voltage will Voltmeter 1 read?

2. What voltage will Voltmeter 2 read?

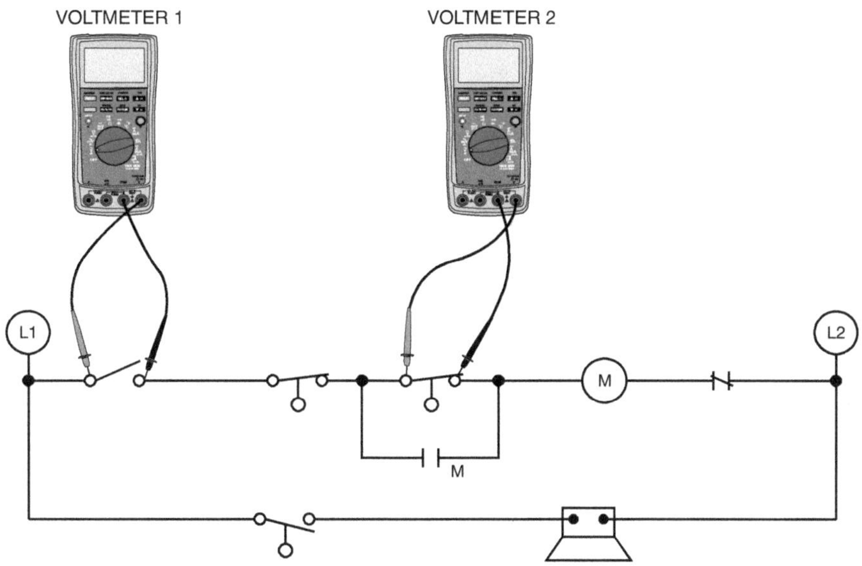

3. What voltage will Voltmeter 1 read?

4. What voltage will Voltmeter 2 read?

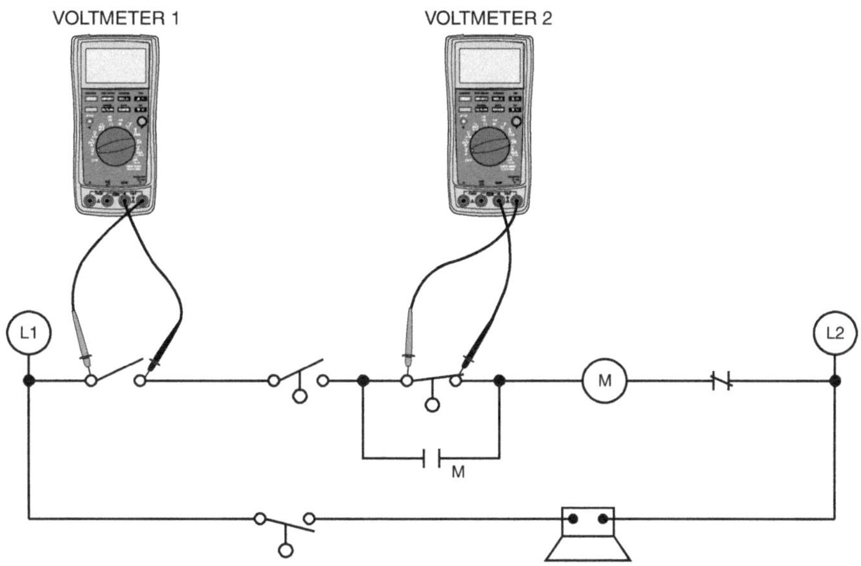

5. What voltage will Voltmeter 1 read?

6. What voltage will Voltmeter 2 read?

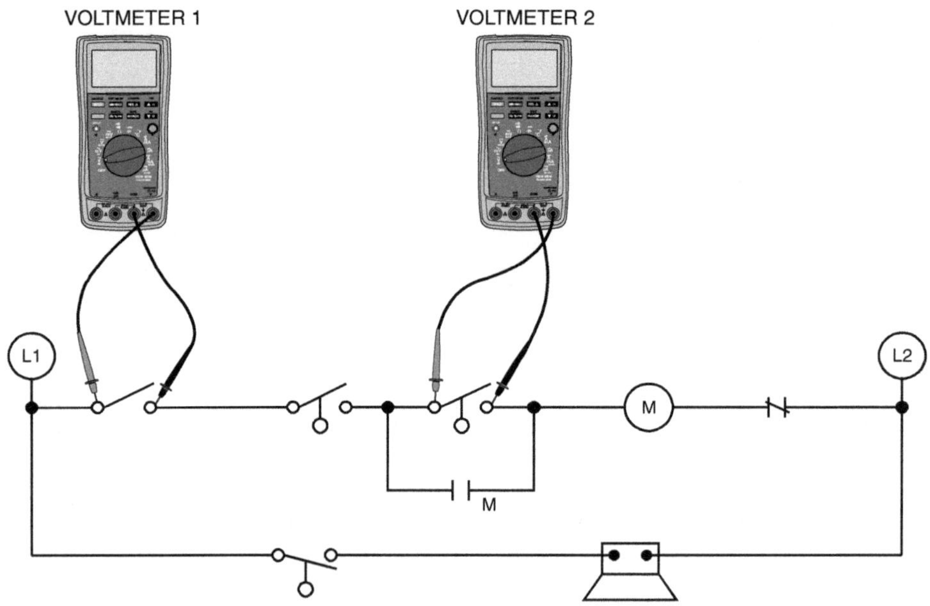

7. What voltage will Voltmeter 1 read?

8. What voltage will Voltmeter 2 read?

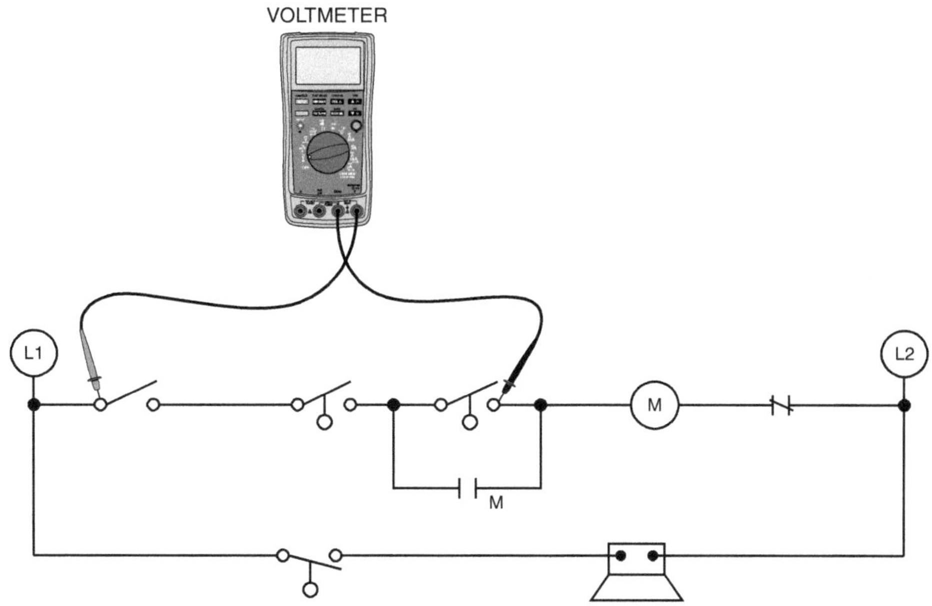

9. What voltage will the voltmeter read?

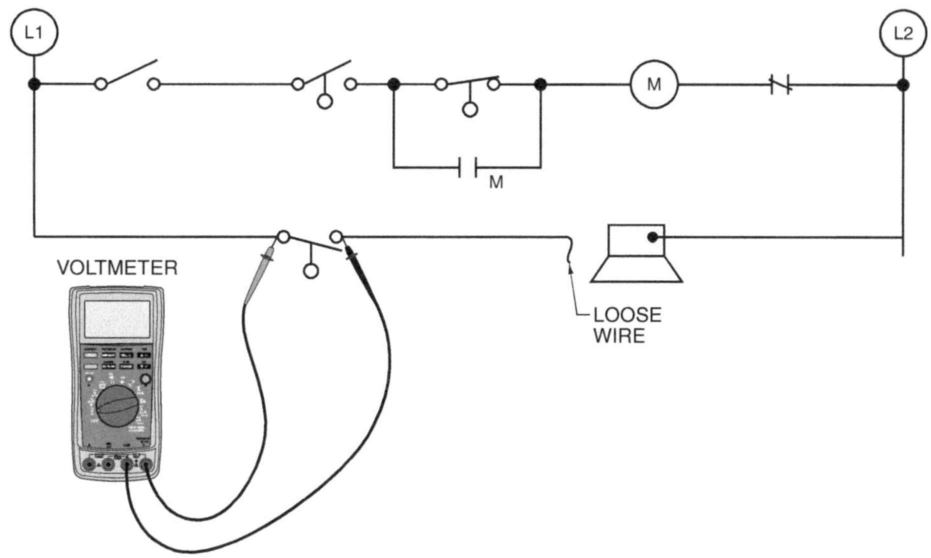

10. What voltage will the voltmeter read?

Activity: Locating Short Circuits

Answer the following questions about short circuits.

1. Why should a fuse blow or a circuit breaker trip open in a circuit with a short in it?

2. What is wrong with a circuit breaker that does not trip open when a circuit develops a short circuit?

3. What could result if the circuit breaker does not trip in this situation?

4. What maintenance activity will help to ensure that breakers will trip when there is a short circuit or a circuit overload?

5. What are visual indications of the location of a short circuit?

6. What safety precautions must be observed before measuring total resistance of a circuit?

7. What position should switches be in to measure total circuit resistance?

8. How could jumper wires be used to close contacts?

9. Briefly describe the process of finding a branch circuit that contains a short circuit.

A line diagram shows several loads and switches connected to a power circuit. The total resistance of the circuit is being monitored with a meter. Portions of the circuit are disconnected one at a time to locate a short.

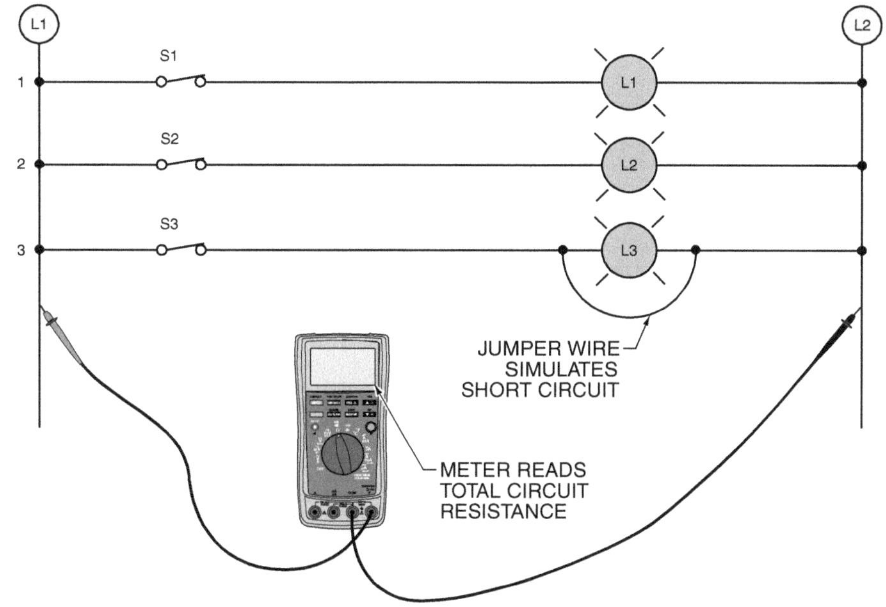

10. What happens to the total resistance if Rung 1 is disconnected from the circuit?

11. What happens to the total resistance if Rung 2 is disconnected from the circuit?

12. What happens to the total resistance if Rung 3 is disconnected from the circuit?

13. What should be done once the short is located?

14. Suggest two conditions that might cause a short circuit.

Activity: Testing Circuit Breakers and Fuses

Answer the following questions about testing circuit breakers and fuses.

1. What visual indication is there that a circuit breaker has tripped?

2. How is a breaker reset?

3. How is a circuit breaker tested using an ohmmeter?

4. What problem could cause a circuit breaker to offer 40 Ω of resistance when it is in the closed position?

5. What ohmmeter readings would be expected if the circuit breaker contacts were melted together so that they would not open?

A fuse can be tested with a voltmeter or a multimeter set to measure voltage.

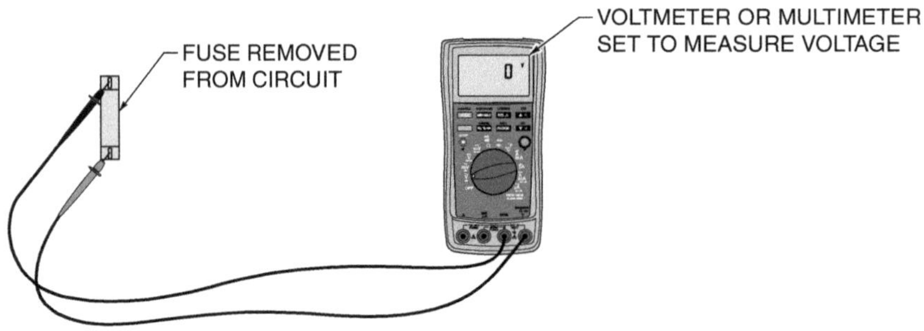

6. What voltage reading would be expected from a good fuse?

7. What voltage reading would be expected from a bad fuse?

A fuse can be tested with an ohmmeter or a multimeter set to measure resistance.

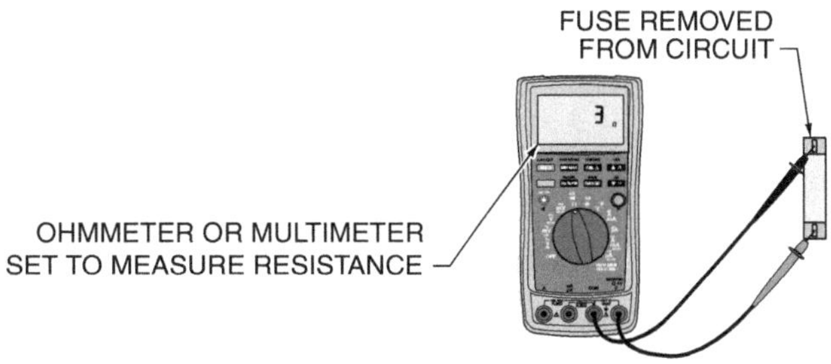

8. What resistance reading would be expected from a good fuse?

9. What resistance reading would be expected from a bad fuse?

10. What ohmmeter reading would be expected from a good time-delay fuse?

Activity: Testing Transformers

The primary and secondary coils of a transformer are tested with voltmeters or multimeters set to measure voltage.

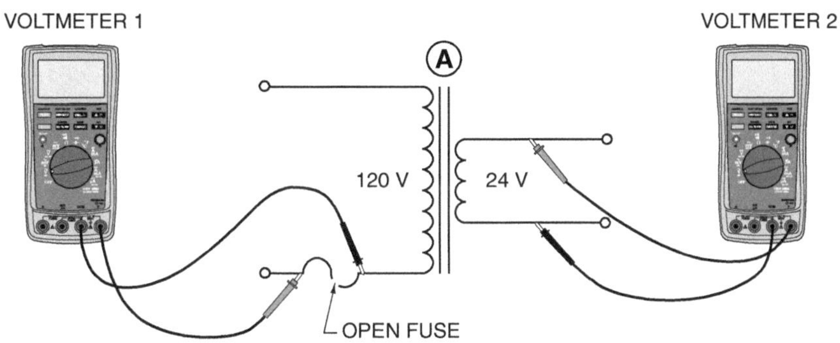

1. What will Voltmeter 1 read for the fuse in Transformer A?

2. What would Voltmeter 1 read if the fuse in the transformer were good?

3. What will Voltmeter 2 read for Transformer A with the blown fuse?

4. What would Voltmeter 2 read if the fuse in the transformer were good?

The primary or secondary coils of several transformers are tested with ohmmeters or multimeters set to measure resistance.

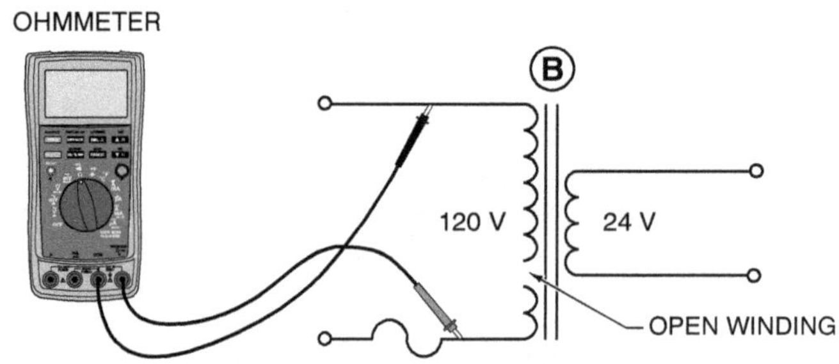

5. What will the ohmmeter read for Transformer B?

6. What would the ohmmeter read for a good transformer?

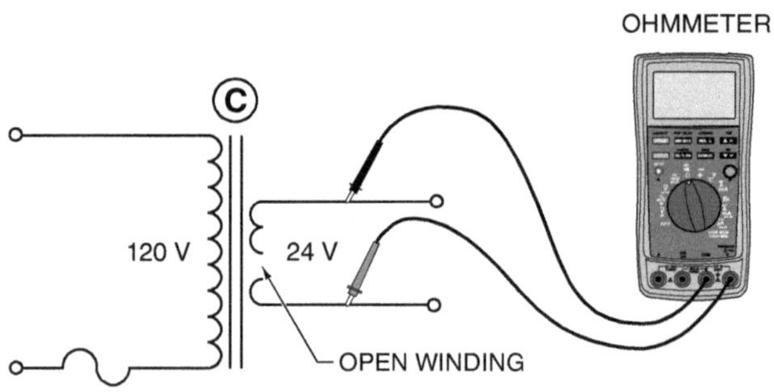

7. What will the ohmmeter read for Transformer C?

8. What would the ohmmeter read for a good transformer?

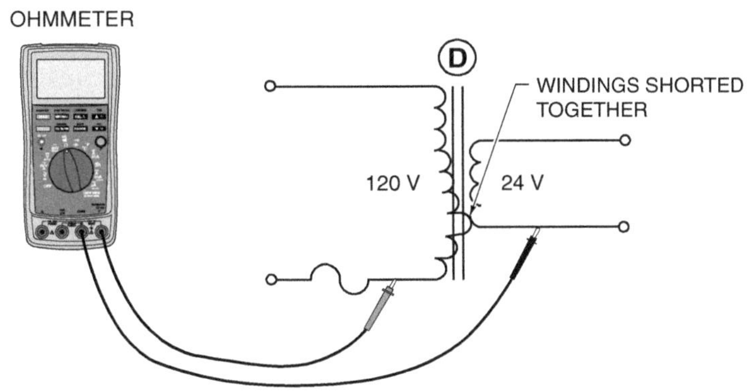

9. What will the ohmmeter read for Transformer D?

10. What would the ohmmeter read for a good transformer?

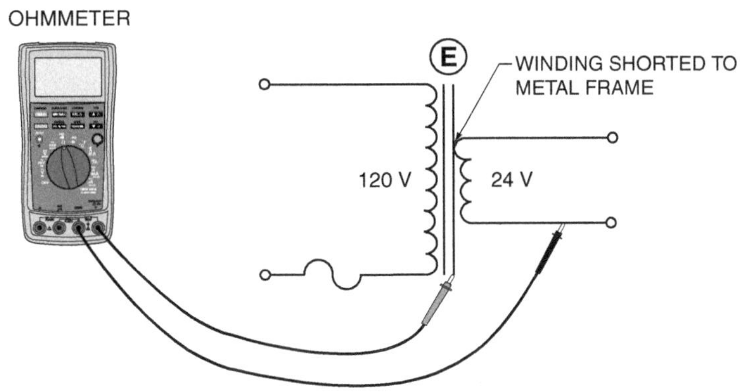

11. What will the ohmmeter read for Transformer E?

12. What would the ohmmeter read for a good transformer?

13. Are transformers repairable?

14. After a transformer has been damaged by overheating, what can be done to prevent the problem from reoccurring on the replacement transformer?

Activity: Testing Coils

When a coil is damaged and develops a short circuit, it usually burns open quickly from the overheating. A voltmeter or multimeter set to measure voltage can be used to detect open coils.

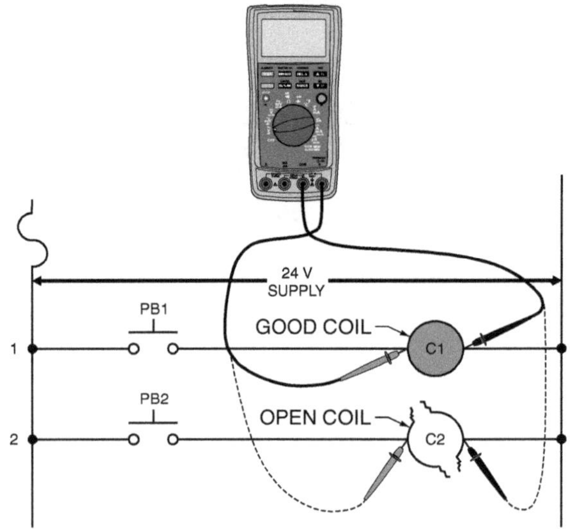

1. What voltage reading would be expected across C1 when PB1 is open?

2. What voltage reading would be expected across C1 when PB1 is closed?

3. What voltage reading would be expected across C2 when PB2 is open?

4. What voltage reading would be expected across C2 when PB2 is closed?

5. What resistance reading would be expected from a good coil?

6. What resistance reading would be expected from an open coil?

Activity: Testing Three-Phase Motors

Before a three-phase motor is replaced, it should be tested to determine if the motor coils are open, shorted together, or shorted to the metal frame of the motor. An ohmmeter or a multimeter set to measure resistance is connected to test each of the motor's coils. Possible readings are no resistance, some resistance, or overload (OL).

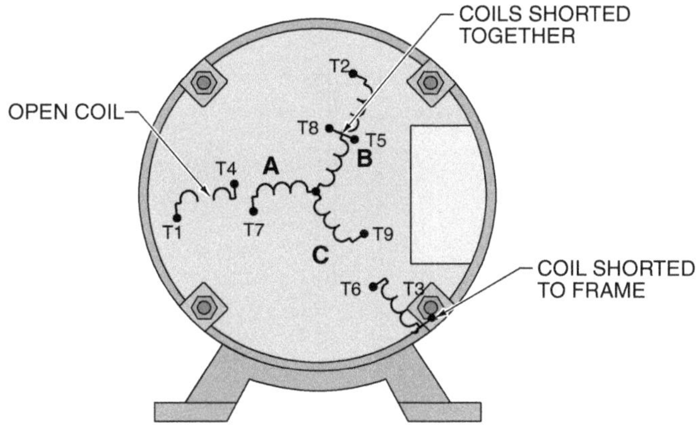

1. What is the resistance between T1 and T4?

2. What is the resistance between T2 and T5?

3. What is the resistance between T3 and T6?

4. What is the resistance between T7 and T9?

5. What is the resistance between T4 and frame?

6. What is the resistance between T5 and frame?

7. What is the resistance between T6 and frame?

8. What is the resistance between T7 and frame?

9. What is the resistance between T1 and T2?

10. What is the resistance between T1 and T6?

11. What is the resistance between T1 and T7?

12. What is the resistance between T2 and T3?

13. What is the resistance between T2 and T7?

Activity: Troubleshooting a Garage Door Opener

A garage door opener found in a commercial application includes a motor to raise and lower the door, a control cabinet, and two limit switches.

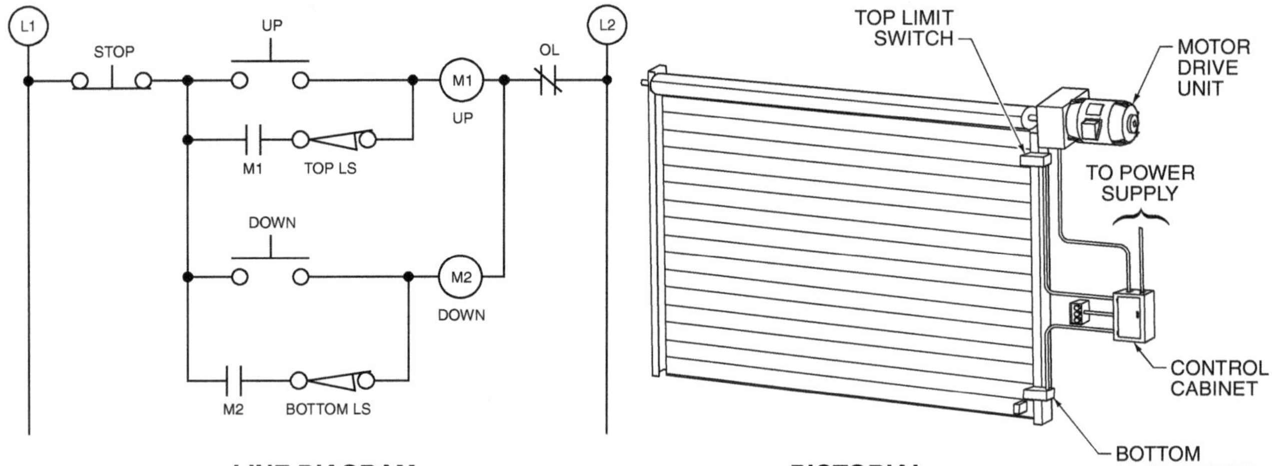

LINE DIAGRAM PICTORIAL

1. If the top limit switch were out of alignment and would not open, what might happen to the motor when the door reached the top of its travel?

2. What mechanical problem could cause the garage door to stop halfway up?

3. How would the electrical circuit work if the bottom limit switch were misaligned and stayed open all the time?

4. Would it be possible to overload the motor if the bottom limit switch were misaligned and stayed closed all the time?

5. What type of mechanical damage to the stop button contacts could prevent the circuit from operating?

6. What electrical problem could result in the motor if the chain drive connecting the motor and garage door were misaligned?

Activity: Troubleshooting a Motor Control Circuit

A motor control circuit that controls three motors from a single start-stop pushbutton station is not operating correctly.

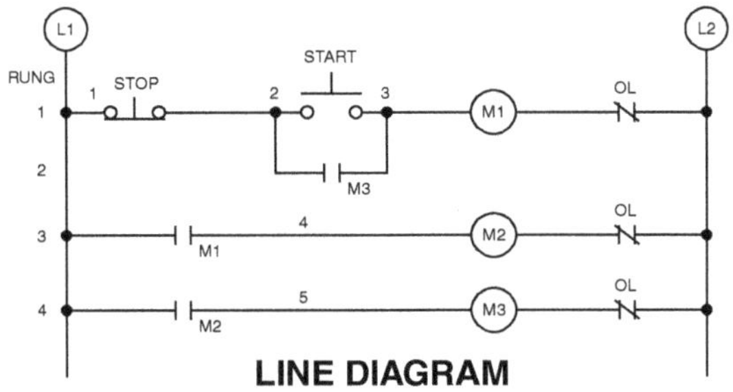

LINE DIAGRAM

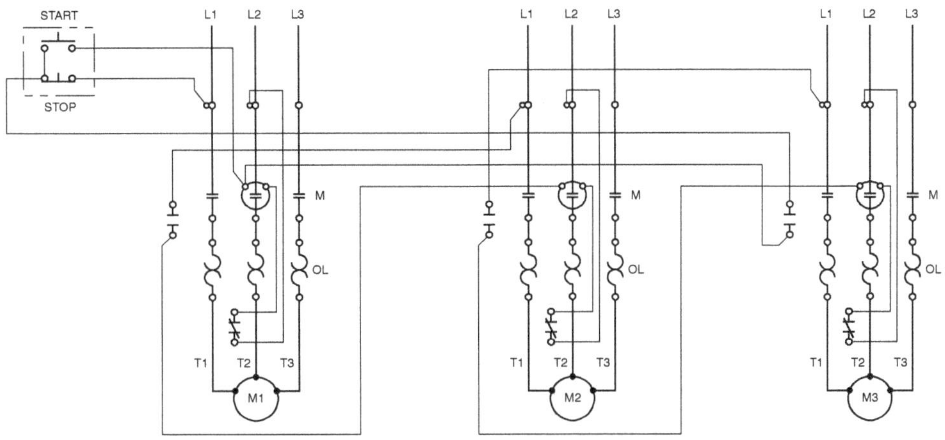

WIRING DIAGRAM

1. List four possible causes for M2 not energizing.

2. How might a loose connection be located?

3. If the M2 coil were open and would not energize, would the M2 contacts on Rung 4 close?

4. What is the likely problem if the fuse blows every time M2 is energized?

Activity: Troubleshooting a Heating Element Control Circuit

An electrical heating element is used to heat air in an HVAC system. A fan directs cool air through the duct and over the heating element, which heats the air. An overheat temperature sensor near the element opens a normally closed switch in the electrical circuit when the temperature around the element exceeds a safe operating level.

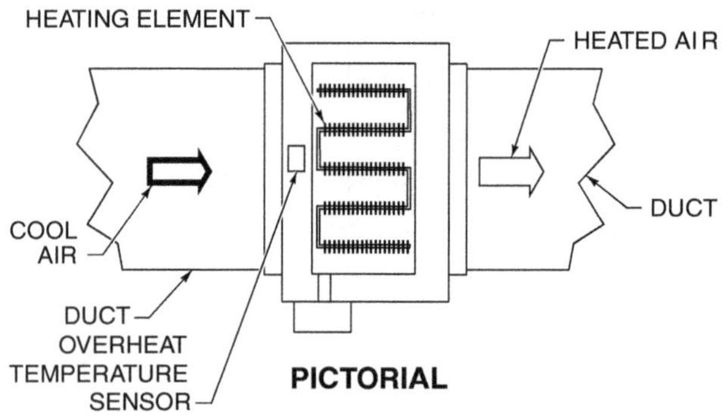

PICTORIAL

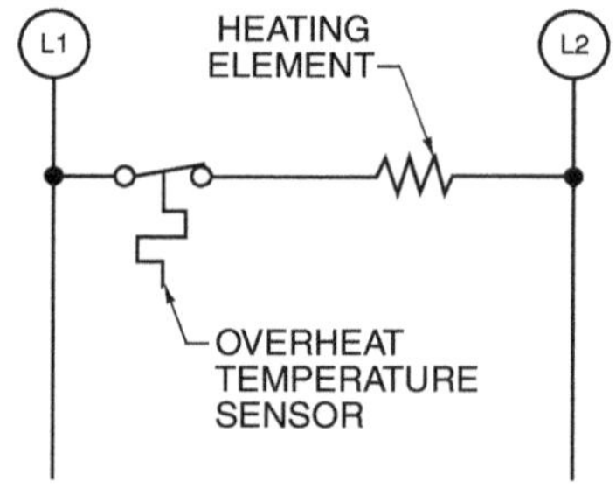

LINE DIAGRAM

1. What might cause the area around the heating element to overheat?

2. If the fan were working properly, what problem could cause a severe reduction in airflow?

3. What resistance reading would indicate that the heating element is bad?

4. If the heating element were open, what would the voltage reading across the heating element be if the overheat temperature sensor were also open?

5. If the heating element overheated and burned open, what additional device must have failed?

6. What preventive maintenance do electric heating elements require?

7. What are the primary safety concerns when working on this type of heating element?

8. If the voltage at the heating element were 0 V, would the fan in the air handler still be operating?

A separate HVAC system uses a set of three heating elements to heat air to a zone.

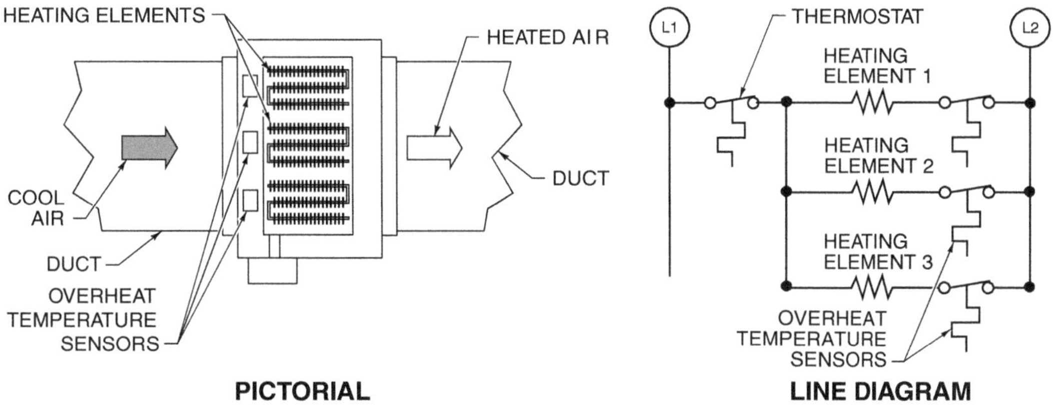

9. If Heating Element 2 were open, would this affect the operation of elements 1 and 3?

SEMICONDUCTORS

SECTION 7.1

REVIEW QUESTIONS

Name ______________________________ **Date** ____________________

True-False

T F **1.** Solid-state devices are extremely reliable but not very fast.

T F **2.** Doping alters the structure of semiconductor material to make the P-type material and N-type material used in solid-state devices.

T F **3.** When free electrons fill holes in P-type material, current flows through the device.

Multiple Choice

______________ **1.** A(n) ___ is a material that has little resistance and permits electrons to move through it easily.
- A. semiconductor
- B. insulator
- C. diode
- D. conductor

______________ **2.** ___-type material is semiconductor material with empty spaces in its crystalline structure.
- A. P
- B. N
- C. PN
- D. NP

______________ **3.** ___-type material is semiconductor material with free electrons in its crystalline structure.
- A. P
- B. N
- C. PN
- D. NP

Completion

______________ **1.** A(n) ___ is an electrically positive space created by a missing electron in P-type semiconductor material.

______________ **2.** A(n) ___ device is an electronic component that switches or controls the flow of current in a circuit with no moving parts.

______________ **3.** A(n) ___ is a material that has a high resistance.

SOLID-STATE DEVICES

SECTION **7.2**

REVIEW QUESTIONS

Name ______________________ **Date** ______________

True-False

(T)	F	**1.** Diodes can be destroyed by heat caused by excessive voltage or current.
T	(F)	**2.** Traffic control lights are often controlled by on-delay timers. Recycle timer function
(T)	F	**3.** An LED is a diode that produces light when current flows through it.
(T)	F	**4.** Diodes are marked or shaped to indicate their anode and cathode.
T	(F)	**5.** A half-wave rectifier is a circuit containing a diode that permits both the positive and negative halves of the AC sine wave to pass.
(T)	F	**6.** An optocoupler is a device that converts electrical input signals to light signals and those light signals back to electrical signals on the output side.

Multiple Choice

______________ **1.** A ___ is a conducting path used to connect components on a PC board.
- A. trace
- (B.) diode
- C. rectifier
- D. triac

______________ **2.** A ___ is an electrical switch that is actuated by a separate circuit.
- A. thermistor
- (B.) relay
- C. transducer
- D. photodiode

______________ **3.** The ___ function is a timer function that, at the start of an input signal, switches the timer contacts immediately and then returns them to their normal state after a set period of time.
- A. recycle
- B. on-delay
- C. off-delay
- (D.) one-shot

______________ **4.** A(n) ___ rectifier is a rectifier containing four diodes.

A. half-wave
B. inverting
C. bridge
D. square

______________ **5.** A ___ switch is a device that reacts to the nearness of a target without physical contact.

A. manual
B. physical
C. proximity
D. thermal

______________ **6.** A ___ is a device that produces electricity when two different metals that are joined together are heated.

A. thermistor
B. transducer
C. triac
D. thermocouple

Completion

______________ **1.** A(n) ___ board is a thin plate of insulating material, such as fiberglass or phenolic, with conducting paths laminated to one or both sides.

______________ **2.** A(n) ___ is an electrical component that converts AC to DC by allowing voltage and current to flow in only one direction.

______________ **3.** A(n) ___ is a device that controls the operation of contacts based on an input signal and a certain time delay.

______________ **4.** A(n) ___ is an electronic component that allows current to pass in only one direction.

______________ **5.** ___ current is the minimum current required to keep an SCR conducting.

______________ **6.** A(n) ___ is a device that combines the transducer effect of a photodiode and the switching capability of a transistor.

ELECTRONIC SYSTEM PRINTREADING

SECTION **7.3**

REVIEW QUESTIONS

Name ______________________________ **Date** ____________________

True-False

T (F) **1.** When an electronic system is represented in detail, it is typically represented with a block diagram.

(T) F **2.** An operational diagram may take any form that conveys the information clearly, such as a list, grid, chart, or graph.

(T) F **3.** The placements of most components on a schematic diagram likely do not match their placements on a PC board.

Multiple Choice

______________ **1.** ___ diagrams are helpful for understanding the relationship of a circuit to nearby circuits and components.

A. Wiring
B. Block
(C). Schematic
D. Operational

______________ **2.** A block diagram is a drawing that illustrates the relationships of the constituent units of a system using ___.

A. dashed lines
B. solid lines
(C) connected blocks
D. lines and symbols

______________ **3.** A(n) ___ diagram is a detailed drawing showing all the components of a system and their interconnections using lines and symbols.

A. schematic
B. block
C. operational
D. electrical wiring

Completion

______________ **1.** A(n) ___ diagram is a drawing that illustrates the operating status of a device under many different conditions.

____________________ **2.** A(n) ___ is used to represent only the inputs, outputs, and connections to other units of a complex unit whose inner workings are not important.

____________________ **3.** A(n) ___ diagram may be similar to an electrical wiring diagram but neglects the physical relationships of the components in favor of grouping by function or connections.

ELECTRONIC SYSTEM MAINTENANCE

SECTION 7.4

REVIEW QUESTIONS

Name ______________________ **Date** ______________

True-False

T (F) **1.** Electronic systems require frequent maintenance and updating.

(T) F **2.** Heat sinks are devices that conduct and dissipate heat away from a component.

(T) F **3.** A common cause of electronic device failures is electrostatic discharge.

Multiple Choice

______________ **1.** ___ is used to promote heat transfer from a component to a heat sink.
- (A.) Thermal grease
- B. Insulation
- C. A fan
- D. A voltage regulator

______________ **2.** A(n) ___ suppressor is a device that reduces random electrical signals on power lines.
- A. voltage
- B. surge
- (C.) noise
- D. overcurrent

______________ **3.** A(n) ___ can be 35,000 V or more, though the total energy released is small.
- A. noise signal
- B. voltage surge
- C. short circuit
- (D.) electrostatic discharge

Completion

noise **1.** Electrical ___ is a disturbance that distorts a voltage and current sine wave.

isolated grounded **2.** A(n) ___ circuit is a circuit that minimizes electrical noise by providing a separate grounding path.

Static electricity **3.** ___ is an electrical charge at rest.

ELECTRONIC SYSTEM TROUBLESHOOTING

SECTION **7.5**

REVIEW QUESTIONS

Name ______________________________ **Date** ______________

True-False

T F **1.** Some voltage is dropped across each diode in a bridge rectifier circuit.

T F **2.** Mechanical contacts should offer nearly 8 V of voltage drop when closed.

T F **3.** A common problem with photoelectric switches includes dirty lenses.

Multiple Choice

______________ **1.** The tone produced by a DMM measuring resistance indicates ___.
- A. reverse bias
- B. voltage drop is within specifications for the diode
- C. low resistance
- D. high resistance

______________ **2.** A typical voltage drop across a good diode is ___ V.
- A. 0.1
- B. 0.6
- C. 2.5
- D. 4.0

______________ **3.** Thermistors are tested by comparing their operation during ___.
- A. a short circuit
- B. power failure
- C. static conditions
- D. changing conditions

Completion

______________ **1.** A quality ___ is used to conduct a basic diode test.

______________ **2.** Thermocouples can be tested using a DMM set to measure ___.

______________ **3.** ___ current is the small amount of current flowing through a solid-state device when it is not conducting.

PROGRAMMABLE LOGIC CONTROLLERS

SECTION **7.6**

REVIEW QUESTIONS

Name ______________________ **Date** ______________

True-False

(T) F 1. A PLC consists of a power supply, input and output modules, and a central processing unit.

(T) F 2. Outputs can be placed in parallel but never in series in a basic PLC circuit.

T (F) 3. During PLC programming, it is important to identify an input as a pushbutton, limit switch, or pressure switch.

Multiple Choice

______ 1. A CPU runs a ___ that is typically created or modified with programming software installed on a computer.
- A. schematic
- B. logic gate
- C. self-check
- (D.) program

______ 2. To create a PLC program, the first step is to ___.
- (A.) create a description of the complete machine operation
- B. determine the sequence of operation
- C. develop a line diagram and legend
- D. indicate the desired logic to the processor

______ 3. A(n) ___ is a solid-state control device that is programmed to automatically control an industrial process or machine.
- A. I/O
- (B.) PLC
- C. CPU
- D. PC

Completion

address 1. A(n) ___ is an identification number assigned to a specific PLC input or output.

SCADA 2. A(n) ___ system is the large-scale integration of equipment, controllers, and data monitoring devices on a communication network.

right 3. Outputs are placed on the ___ side of the circuit in a basic PLC circuit.

PROGRAMMABLE LOGIC CONTROLLER MAINTENANCE

SECTION 7.7

REVIEW QUESTIONS

Name ______________________________ Date ______________

True-False

T (F) 1. PLC programs are quick and easy to develop and require little maintenance.

(T) F 2. Network-enabled data systems may be able to run backups over the network to a connected server.

Multiple Choice

______________ 1. Preferably, data should be backed up regularly ___.
- A. until the project is completed
- B. in hard copy form only
- C. to a server in the same location
- (D.) and stored in a separate location

______________ 2. It is recommended to save copies of data ___ in case a previous version is needed later.
- A. as hard copies only
- (B.) at multiple stages of program development
- C. for one week
- D. only once

Completion

Cool 1. PLCs should be kept clean, dry, ___, and protected from damage.

Backing up 2. ___ is the process of saving computer data to more than one storage device.

PROGRAMMABLE LOGIC CONTROLLER TROUBLESHOOTING

SECTION **7.8**

REVIEW QUESTIONS

Name ______________________________ **Date** ______________

True-False

(T) F **1.** For PLCs, a test mode may be available that prevents output devices from energizing but still uses the output status indicators.

(T) F **2.** Forcing can be done with the system operating, in which case the loads will deenergize.

T (F) **3.** An input module should be replaced when the input circuit is operating correctly but the input module status indicator does not light.

Multiple Choice

______________ **1.** A ___ is the malfunction of an entire machine.
- (A). total shutdown
- B. partial fault
- C. disabling
- D. software shutdown

______________ **2.** If after a total shutdown the power status indicator is on but the PC Run status indicator is off and the CPU Fault status indicator is on, the PLC is ___.
- A. disabled
- B. disconnected from the power source
- (C). receiving a faulty input
- D. receiving the correct input

______________ **3.** ___ is the prevention of the activation of a PLC output.
- A. A partial fault
- B. Lockout
- C. Forcing
- (D). Disabling

Completion

status indicators **1.** ___ are used to ensure that inputs and outputs are behaving as expected according to the program.

forcing **2.** ___ is the manual activation of a PLC input or output by a command entered through a programming terminal.

Partial fault **3.** A(n) ___ fault is the malfunctioning of only a section or several sections of a machine.

ELECTRONIC SYSTEMS

Chapter 7

ACTIVITIES

Name ______________________________ **Date** ______________

Isolated Receptacle Grounding

______________ **1.** Neutral conductor

______________ **2.** Hot conductor

______________ **3.** Isolated ground conductor

______________ **4.** Building ground for conduit and other equipment

______________ **5.** Identification of isolated grounded receptacles

______________ **6.** Conduit

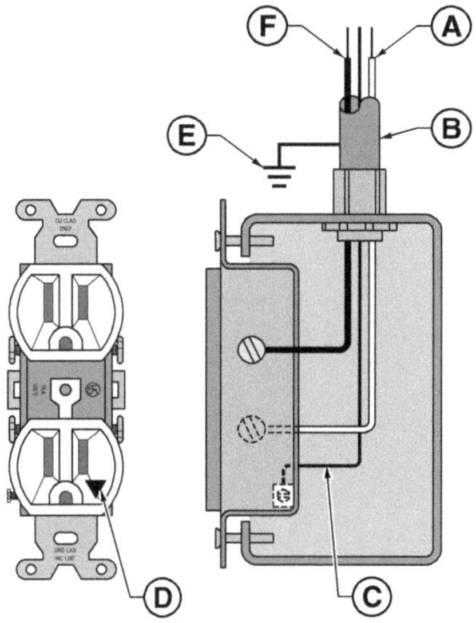

PC Boards

______________ **1.** Terminal contacts

______________ **2.** Electrical components

______________ **3.** Insulated board

______________ **4.** Pads

______________ **5.** Connector

______________ **6.** Bus

______________ **7.** Traces

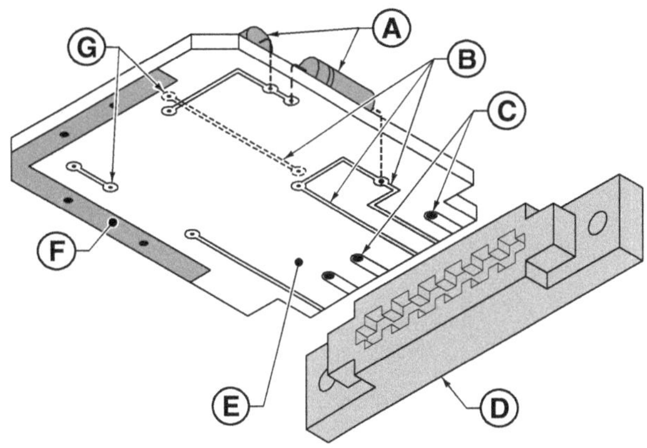

Testing Power Supplies

______________ 1. Measure power supply output

______________ 2. Measure transformer primary and secondary voltage

______________ 3. Measure power supply fuse

______________ 4. Measure voltage source

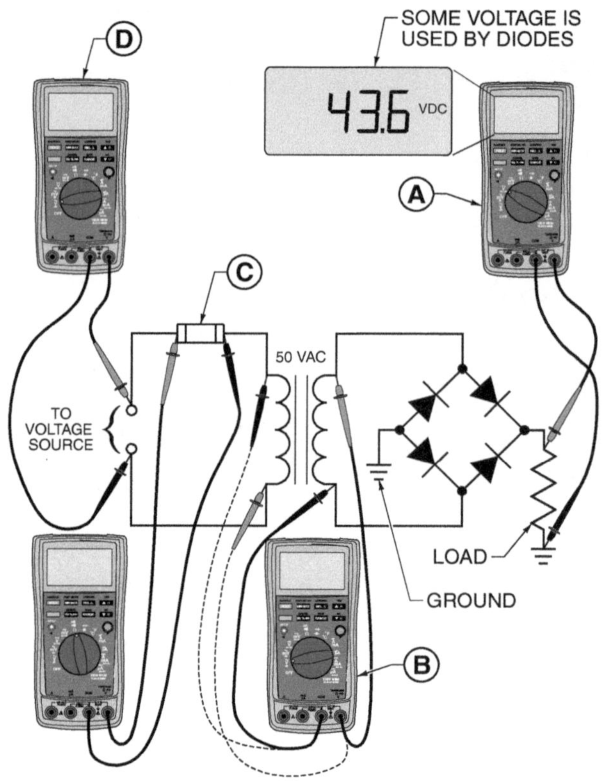

Timer Functions

______________ 1. Recycle

______________ 2. Off-delay

______________ 3. One-shot

______________ 4. Input signal

______________ 5. On-delay

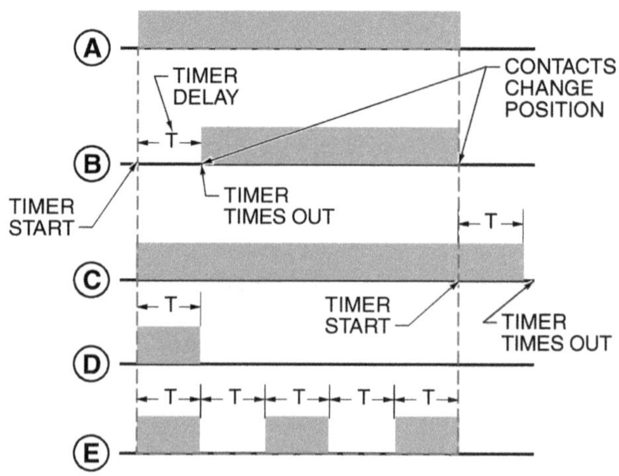

Activity: Mechanical and Electronic Contacts

Mechanical and/or electronic contacts are used in various circuits to switch loads.

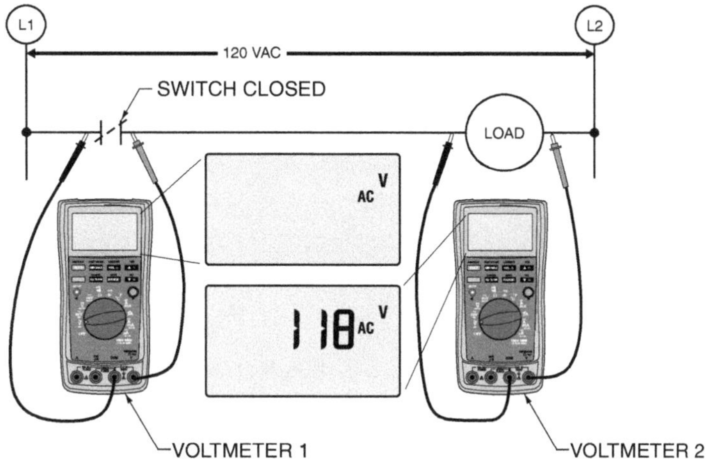

1. How much voltage drop should Voltmeter 1 read across the closed mechanical contact?

2. Is the voltage drop to the load that is indicated by Voltmeter 2 within specifications?

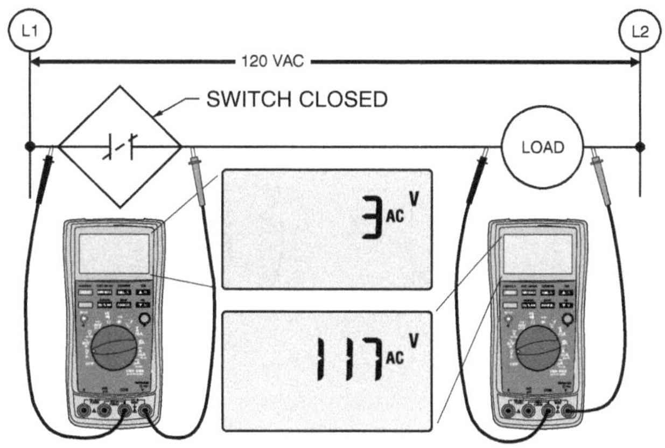

3. Why is there a 3 V drop across this electronic contact in the closed position?

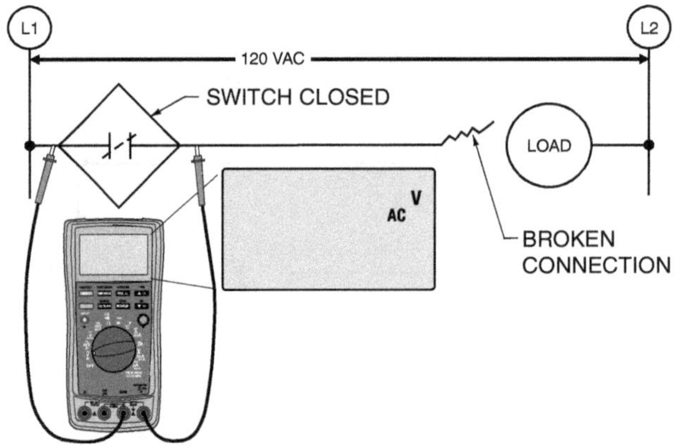

4. How much voltage drop would be indicated across the electronic contact?

5. Would this load operate properly?

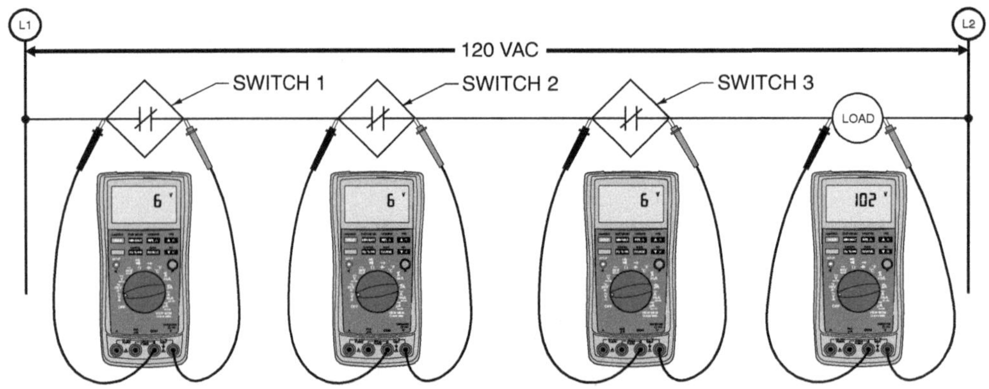

Activity: Testing Diodes

A power supply with a diode is obviously damaged. There are signs of overheating, and the diode is cracked. The diode is a solder-on type.

1. Why is it unlikely that the diode would be replaced by soldering in the field?

2. What three conditions could cause the diode to fail?

Several other diodes on the power supply are tested with a digital multimeter. The first reading tests the diode in forward bias, and the second reading tests the diode in reverse bias.

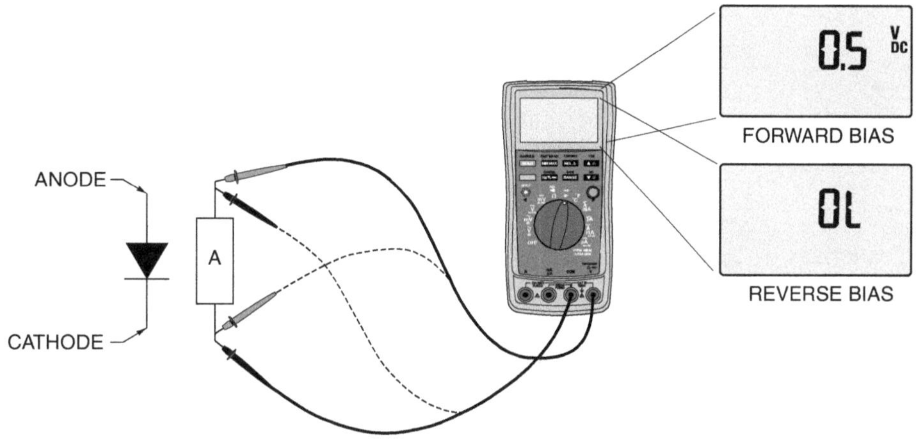

3. Is diode A good?

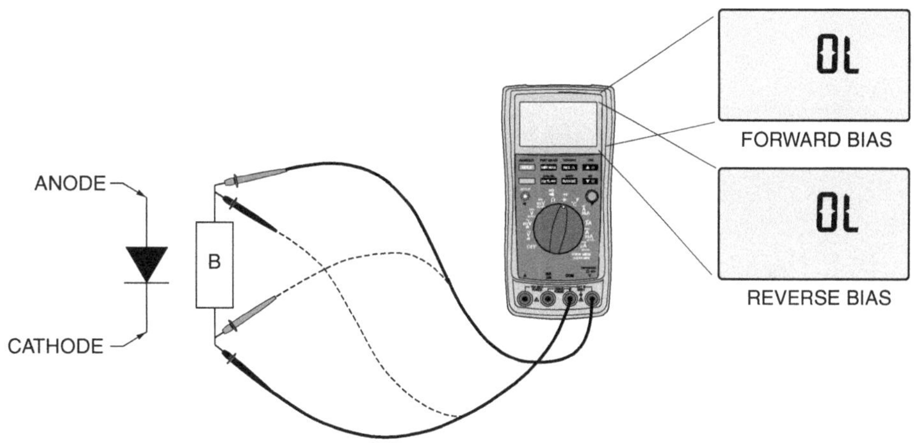

4. Is diode B good?

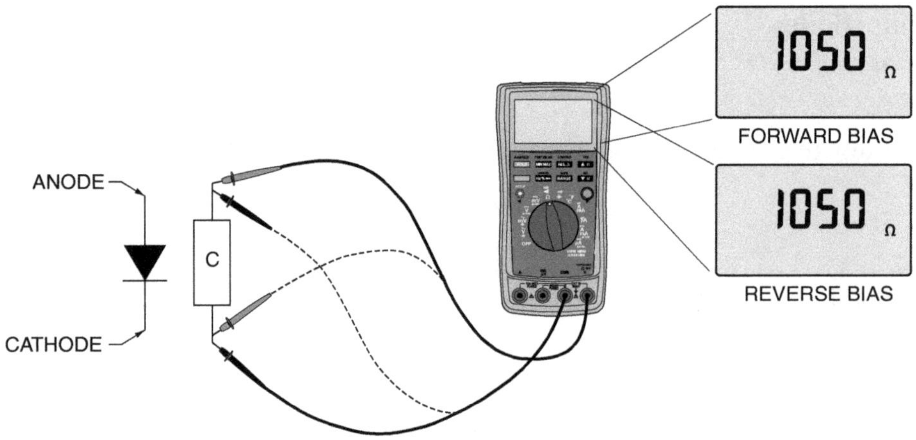

5. Is diode C good?

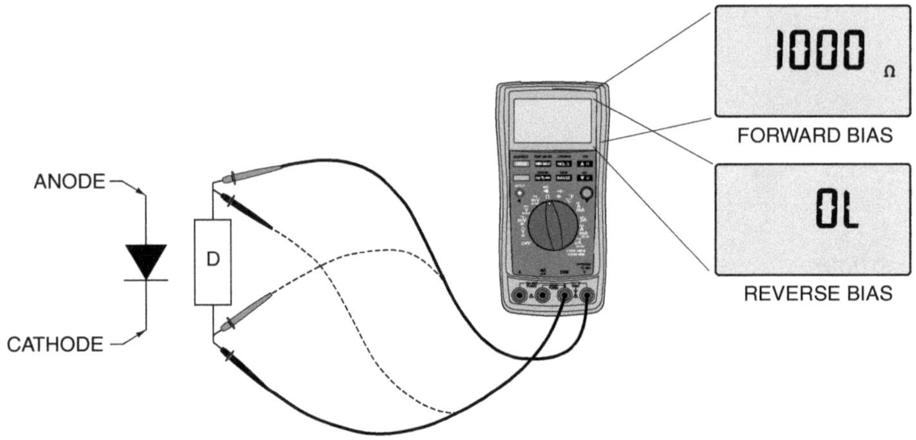

6. Is diode D good?

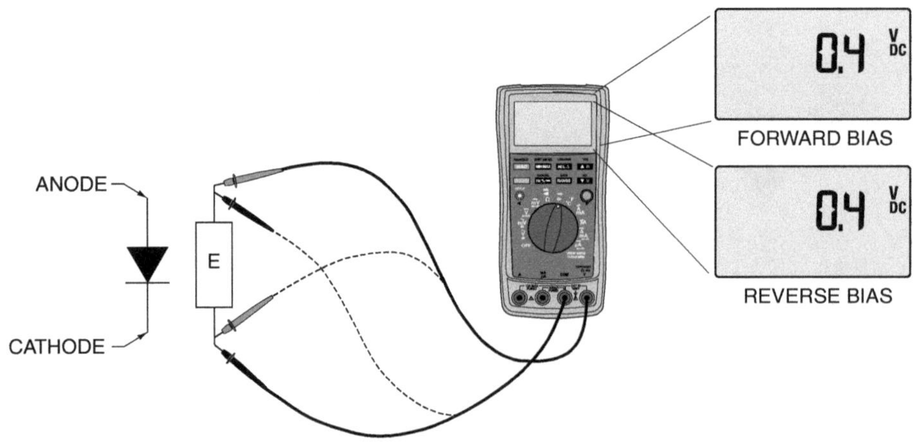

7. Is diode E good?

Activity: Rectifiers

Answer the following questions based on the rectifier circuits.

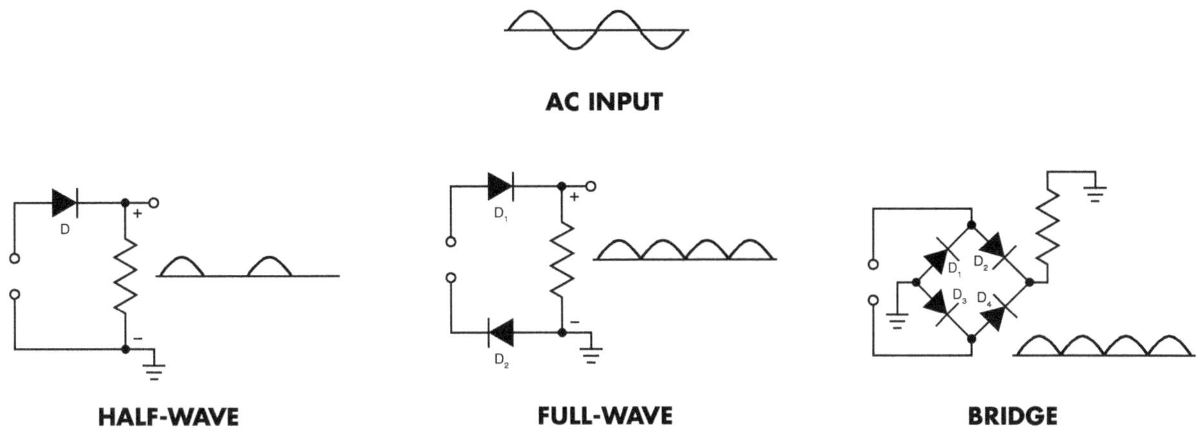

1. Which type of rectifier is the most efficient?

2. Which type of rectifier provides the least amount of power to the output?

A silicone-controlled rectifier (SCR) is used in the power circuit of a DC motor.

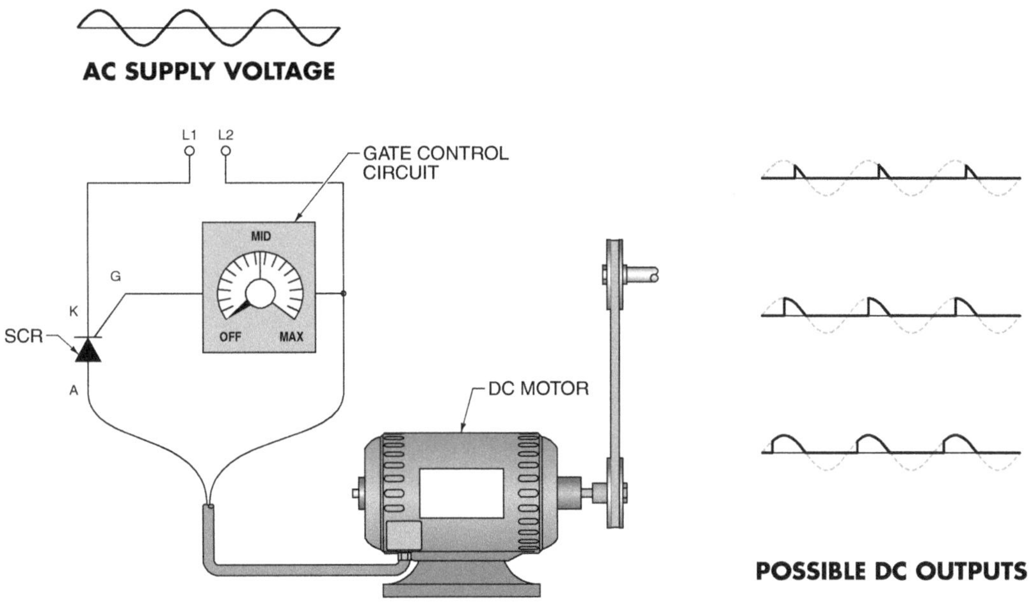

3. What does the SCR do to the voltage supplied to the DC motor?

4. What effect does this have on the motor?

Activity: Triacs

A triac is used in the power circuit for a heating unit installed in an air duct.

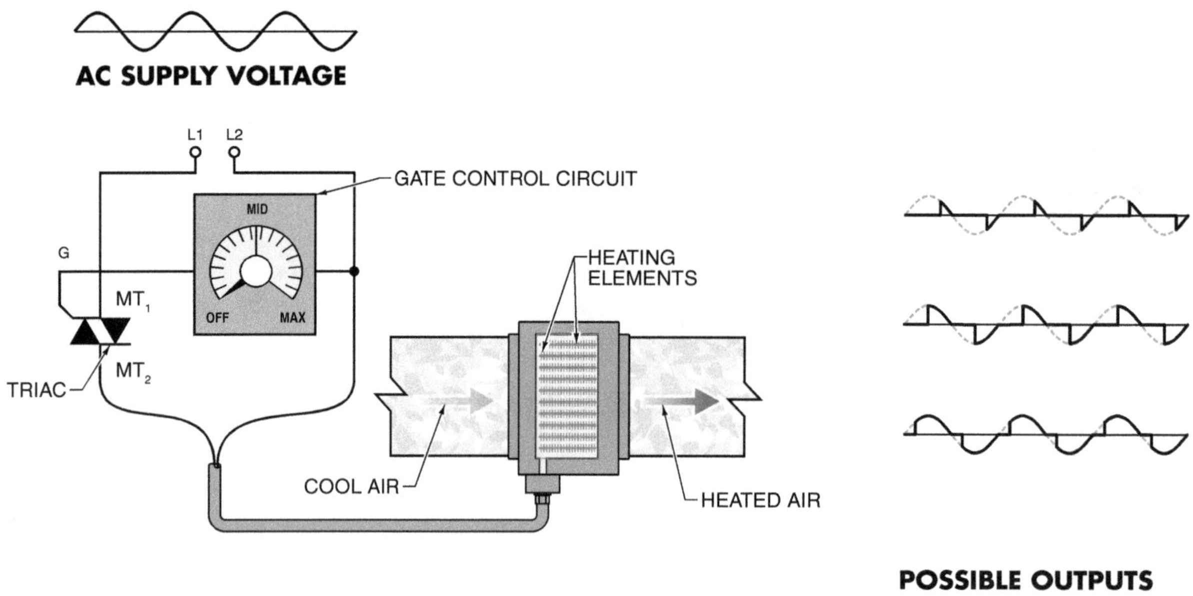

1. Do the heating elements operate on DC power or AC power?

2. What does the triac do to the AC waveform?

3. What is the effect of the output waveforms on the heating elements?

4. What part of the circuit determines the level of output power?

Activity: Testing Thermistors

Thermistors change resistance in response to a change in temperature. Thermistors are tested by measuring their resistance (or current) with a multimeter.

At 85°F, Thermistor A and Thermistor B should have a resistance of 25 Ω ±2%.

1. When testing a thermistor, what technical information is needed?

2. Is Thermistor A good?

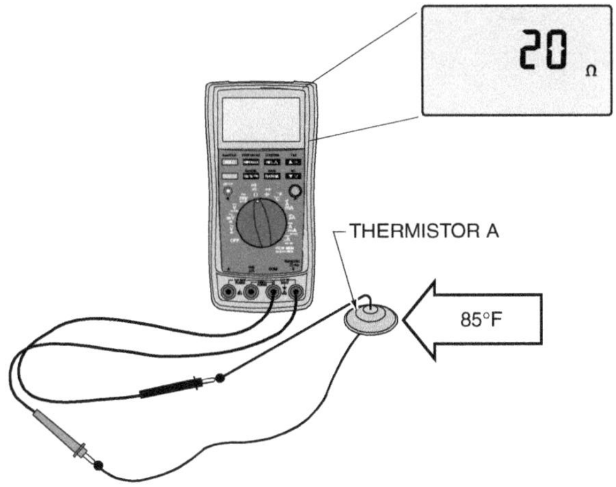

3. What will Thermistor B read?

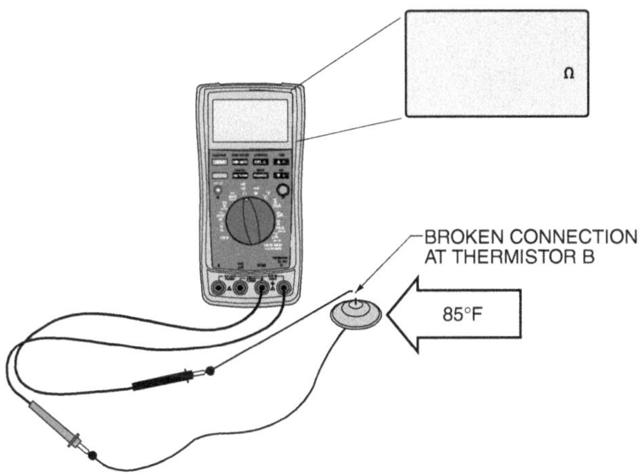

At 75°F, Thermistor C should have a resistance of 50 Ω ±2%. At 100°F, it should have a resistance of 25 Ω ±2%.

4. Will the current flowing through Thermistor C go up or down when temperature increases?

5. What four conditions could cause a thermistor to fail?

Activity: Testing Thermocouples

Thermocouples produce small amounts of electricity when heated. The voltage varies in response to the temperature. Thermocouples are tested by measuring their voltage with a multimeter. At 75°F, Thermocouple A and Thermocouple B should generate 25 mVDC ±2%.

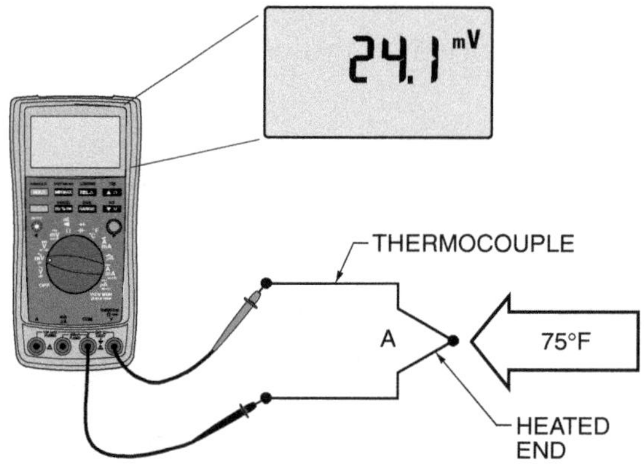

1. Is Thermocouple A good?

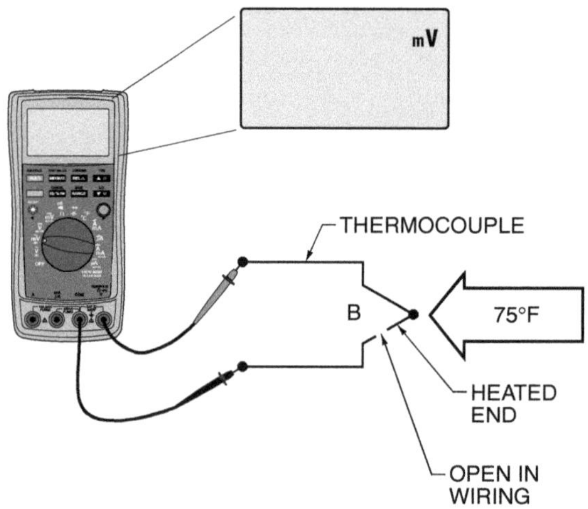

2. How much voltage will Thermocouple B indicate?

3. What would Thermocouple B read if tested by an ohmmeter?

4. Should a good thermocouple read any resistance if tested by an ohmmeter?

At 75°F, Thermocouple C and Thermocouple D should generate 25 mVDC ±2%. At 100°F, Thermocouple C and Thermocouple D should generate 50 mVDC ±2%.

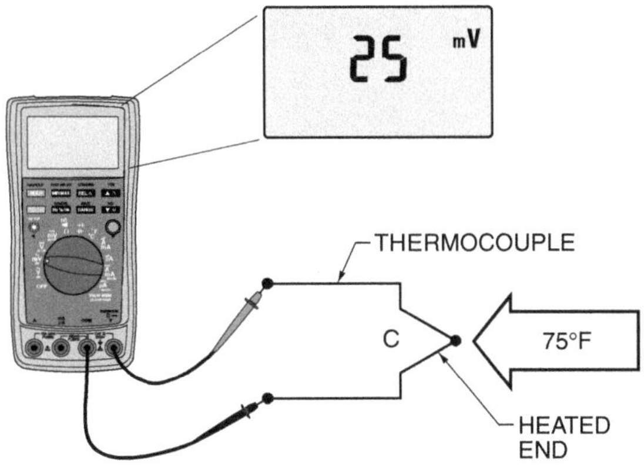

5. Is Thermocouple C good?

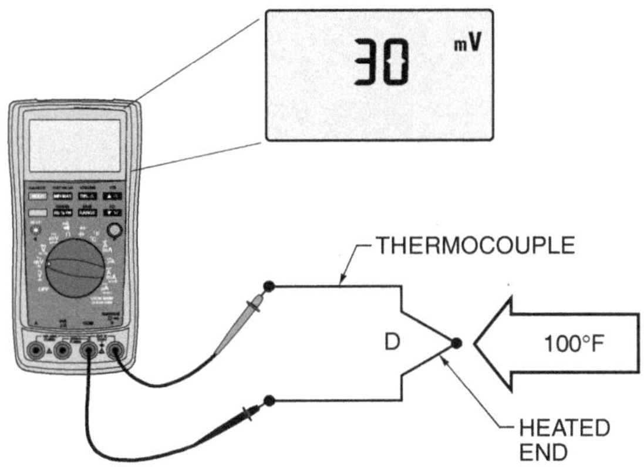

6. Is Thermocouple D good?

7. How can inventory management help minimize downtime when working on equipment controlled by thermocouples?

8. What might happen if a damaged thermocouple in a piece of equipment were replaced with a thermocouple with different specifications?

A piece of equipment is controlled by a system that includes a thermocouple. The thermocouple connection wires are loose.

9. What might happen to the signal from the thermocouple?

10. How might this affect the operation of the equipment?

PLC Inputs and Outputs

A control circuit uses switches (S1–S6) to activate loads (L1–L4) based on the logic programmed into a PLC. This logic is shown in both a line diagram and a PLC program diagram. The switches and loads are currently in their normal states.

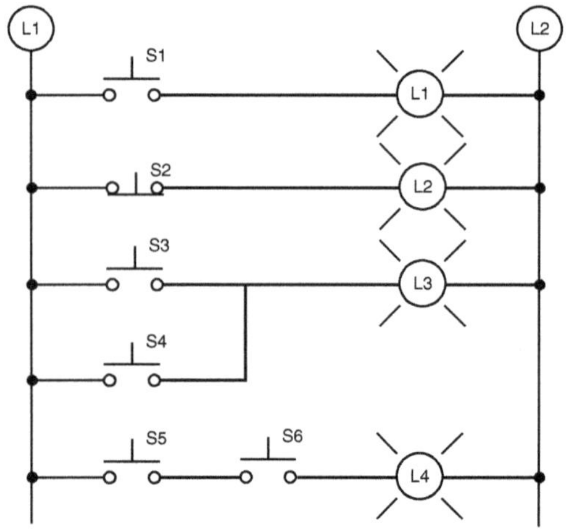

LINE DIAGRAM

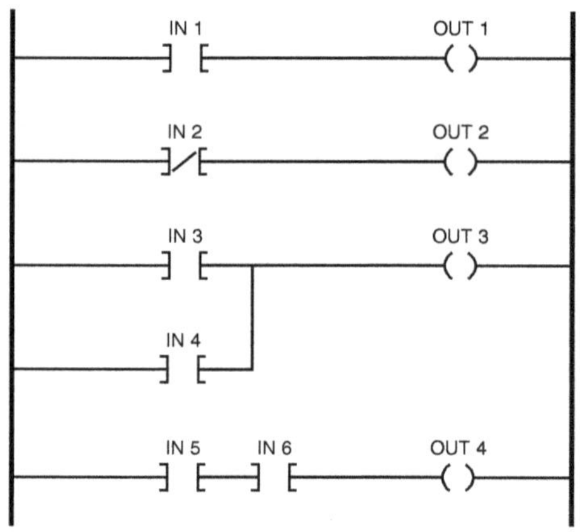

PLC DIAGRAM

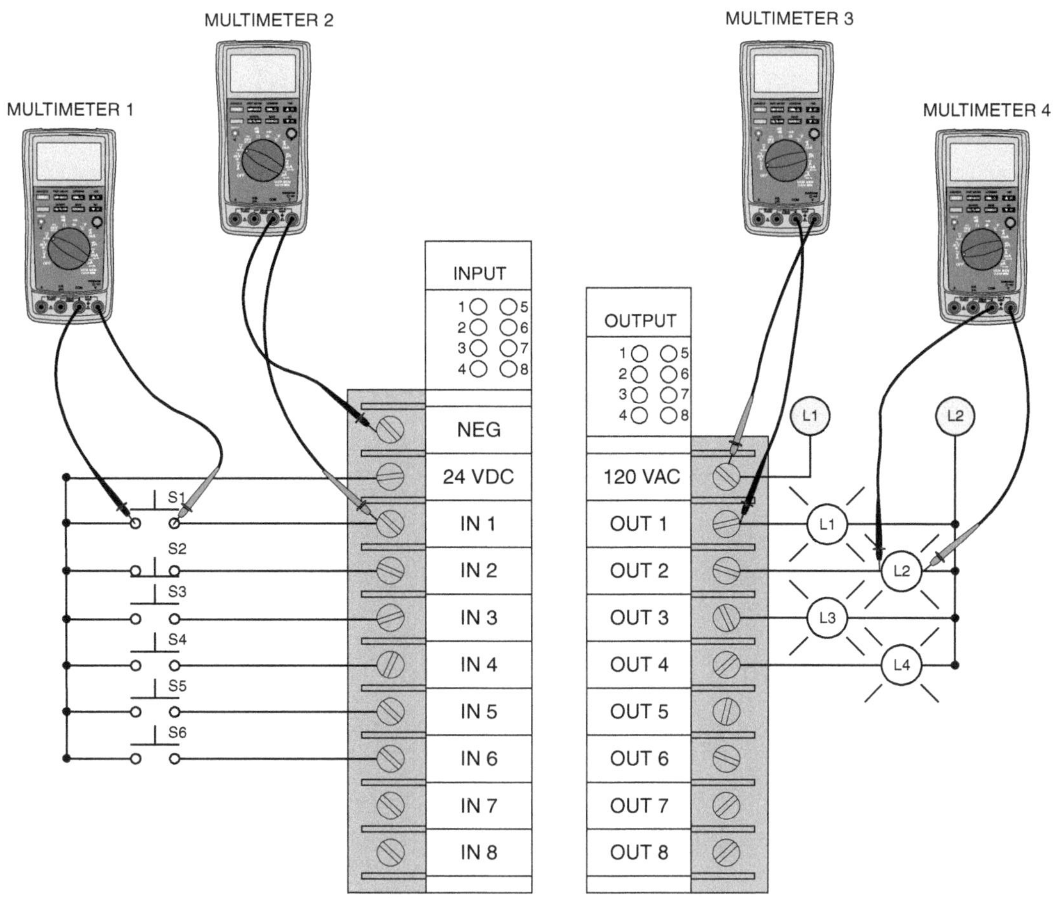

1. Which input and output indicator light(s) should be illuminated?

2. What reading should Multimeter 1 indicate?

3. What reading should Multimeter 2 indicate?

4. What reading should Multimeter 3 indicate?

5. What reading should Multimeter 4 indicate?

Switch 1 (S1) is now activated.

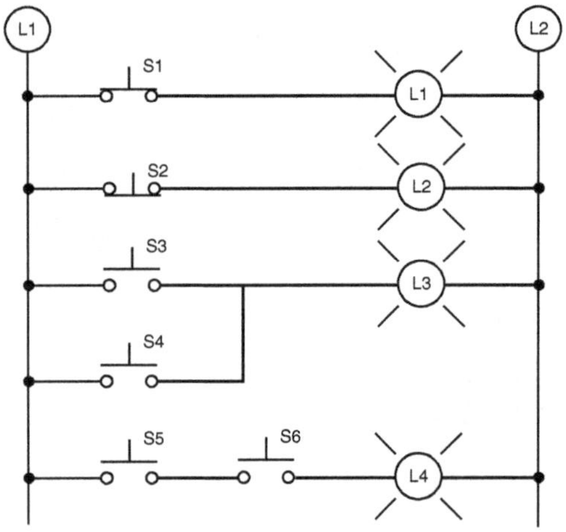

LINE DIAGRAM

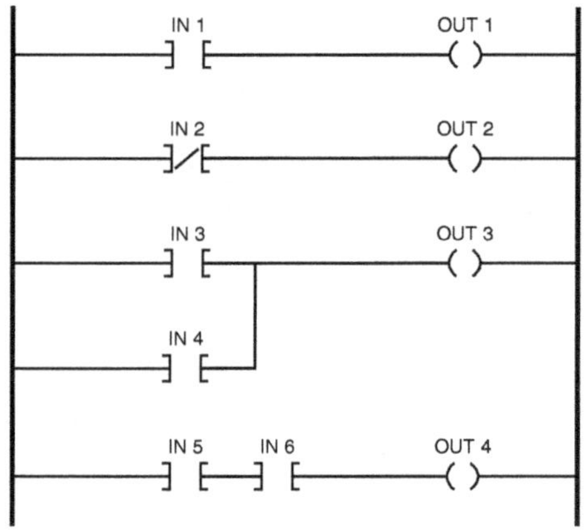

PLC DIAGRAM

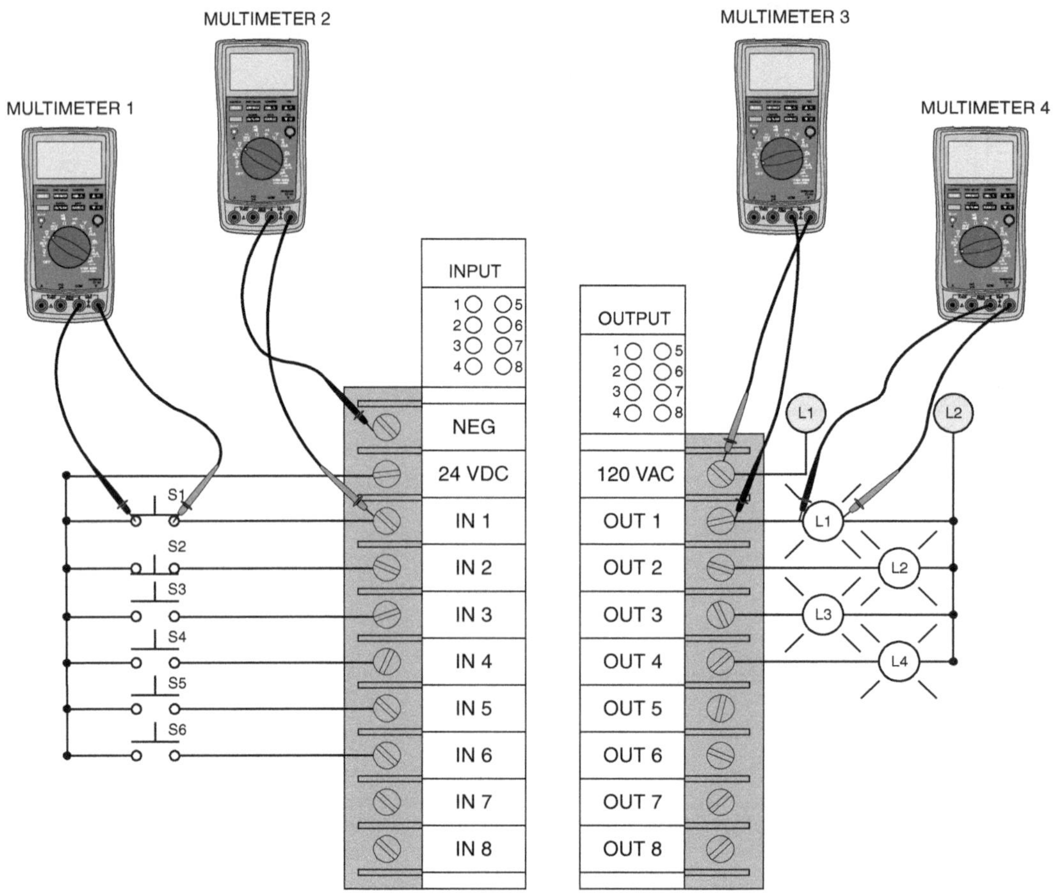

6. Which input and output indicator light(s) should be illuminated?

7. What reading should Multimeter 1 indicate?

8. What reading should Multimeter 2 indicate?

9. What reading should Multimeter 3 indicate?

10. What reading should Multimeter 4 indicate?

Switch 1 (S1) is deactivated, and Switch 2 (S2) is activated.

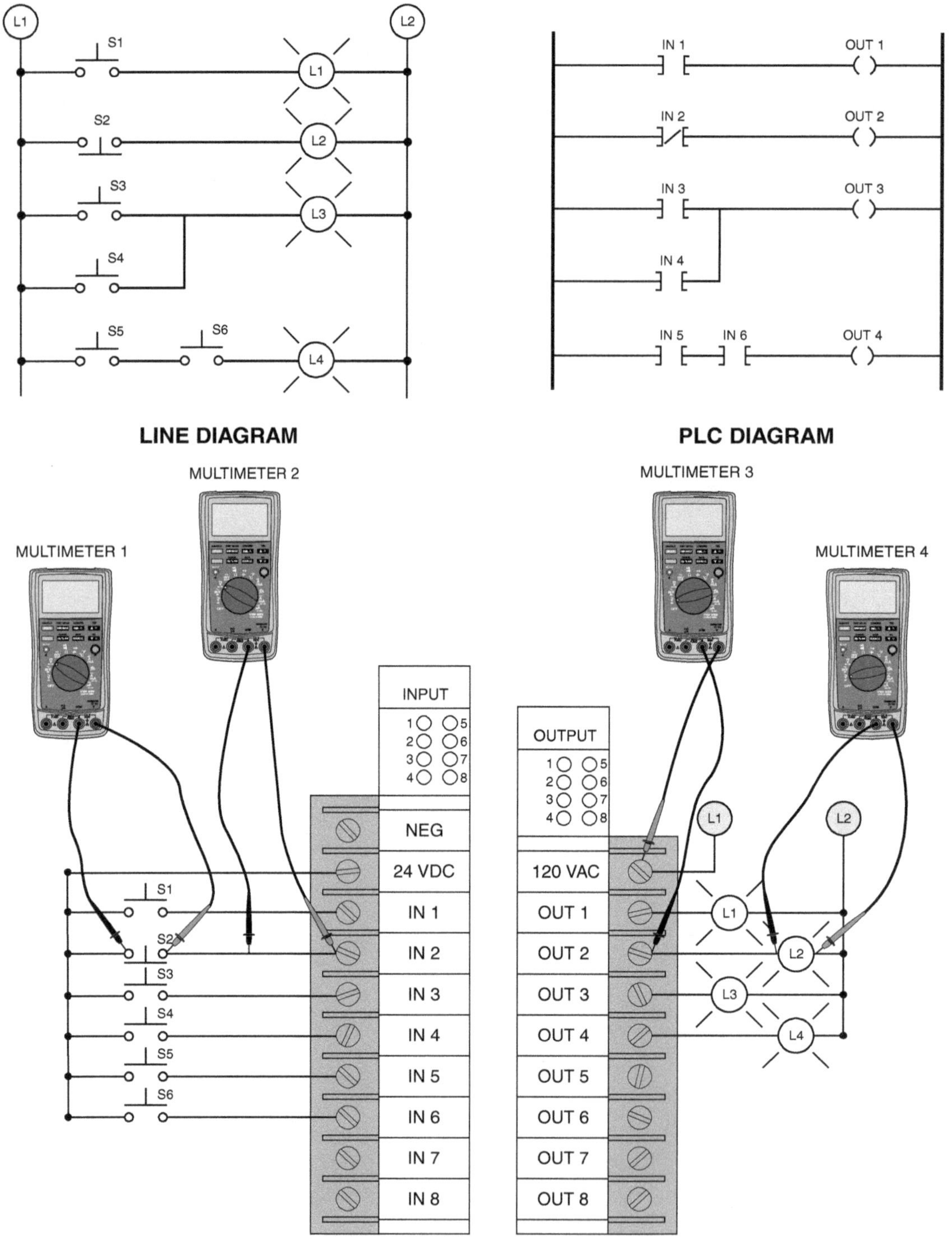

11. Which input and output indicator light(s) should be illuminated?

12. What reading should Multimeter 1 indicate?

13. What reading should Multimeter 2 indicate?

14. What reading should Multimeter 3 indicate?

15. What reading should Multimeter 4 indicate?

Switch 2 (S2) is deactivated, and Switch 3 (S3) is activated.

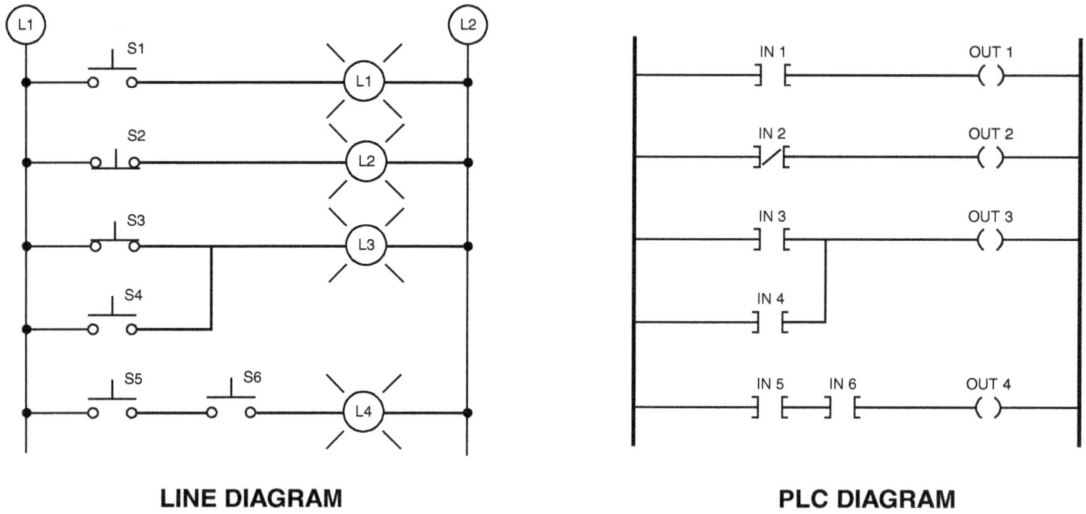

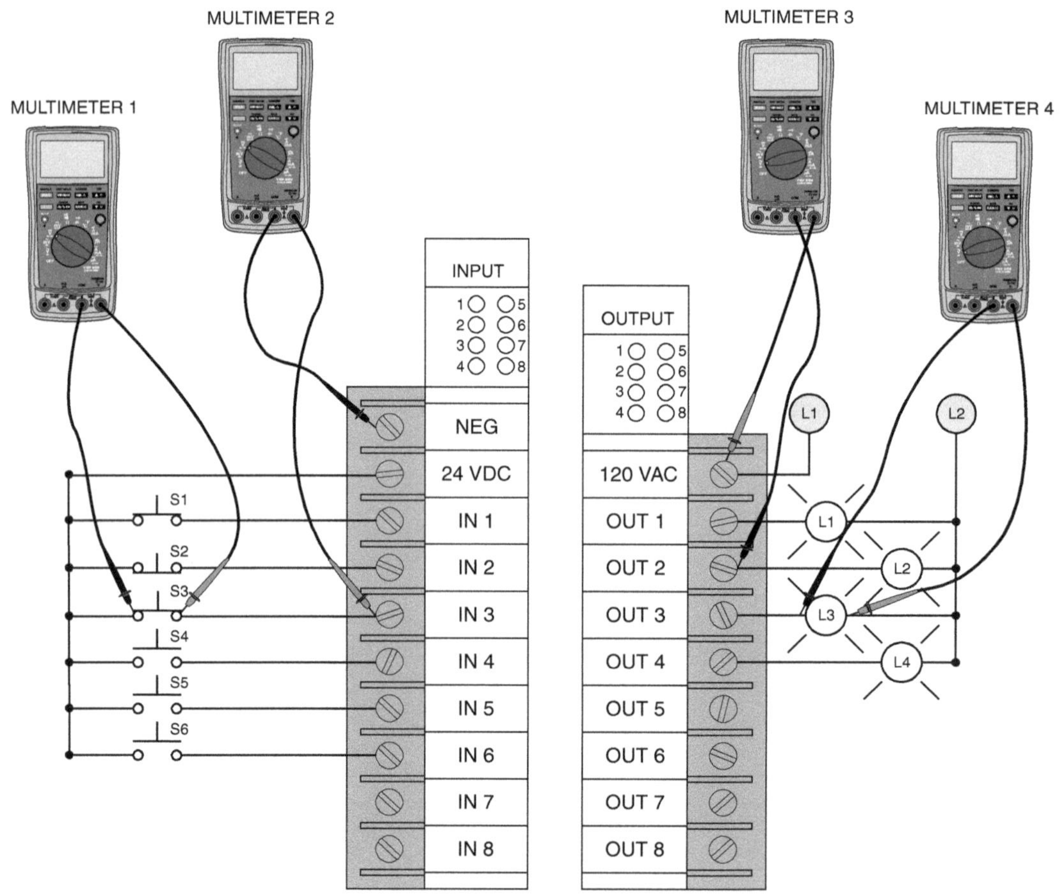

16. Which input and output indicator light(s) should be illuminated?

17. What reading should Multimeter 1 indicate?

18. What reading should Multimeter 2 indicate?

19. What reading should Multimeter 3 indicate?

20. What reading should Multimeter 4 indicate?

Activity: PLC Troubleshooting

A PLC-controlled circuit is not operating as expected. The indicator lights on the input and output modules are used to troubleshoot the problem.

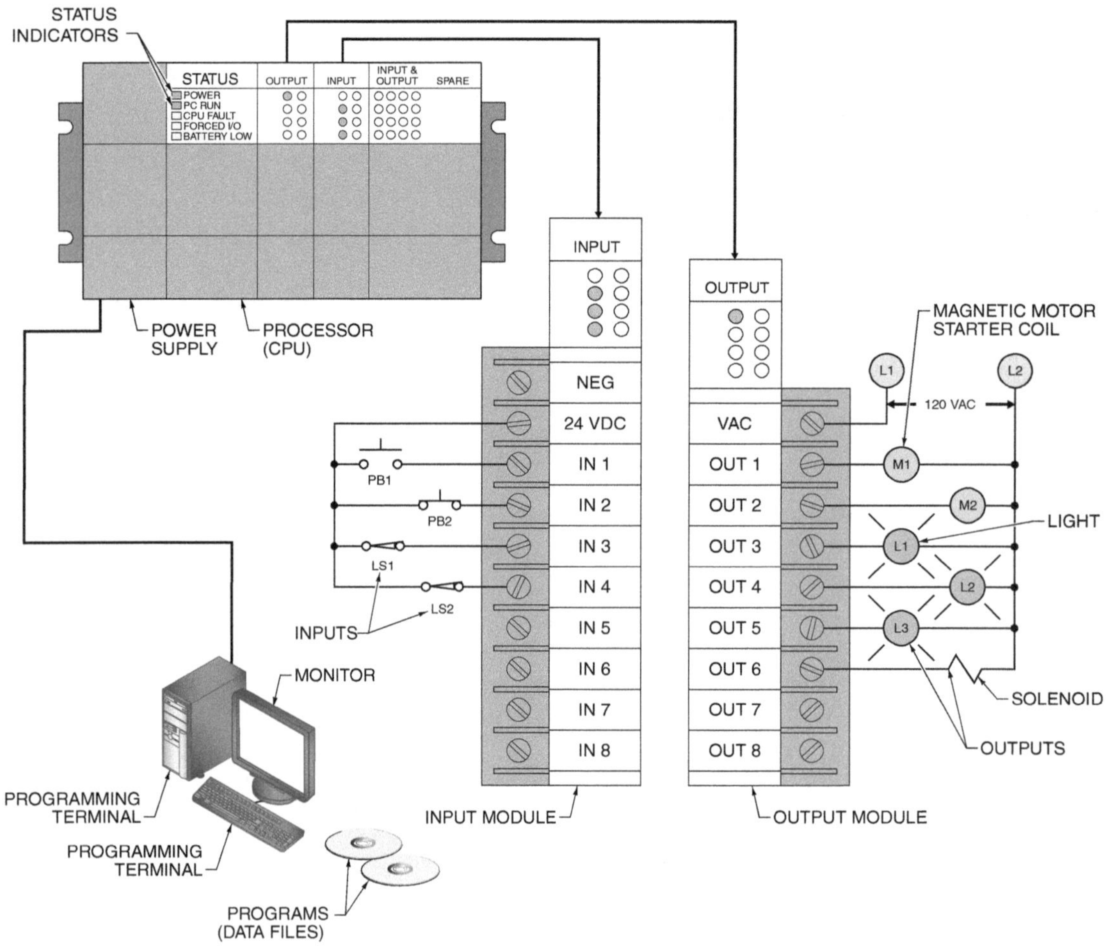

1. What might be the problem if PB1 is pressed but Input 1 does not light?

2. What might be the problem if Input 1 stays lit when PB1 is released?

3. What might be the problem if PB1 is pressed and Input 1 lights but Output 1 does not light as expected?

4. What might be the problem if PB1 is pressed and Output 6 lights but the solenoid does not energize?

5. What can be done if the program for the PLC is unintentionally altered and operating erratically?

6. If a load works when it is forced through the PLC program but does not work when the inputs are operated, what is one likely cause of the problem?

REFRIGERATION PRINCIPLES

SECTION **8.1**

REVIEW QUESTIONS

Name ______________________________ **Date** ____________________

True-False

T	F	**1.** The most common method of moving heat with refrigerants is by evaporation.
T	F	**2.** A condenser is a mechanical device that compresses gas.
T	F	**3.** In mechanical compression refrigeration systems, the high pressure may also be known as suction pressure and the low pressure may also be known as head pressure.
T	F	**4.** Heat flows only from a warmer temperature to a cooler temperature.
T	F	**5.** All substances exist in either a solid, liquid, or gaseous state.

Multiple Choice

______________ **1.** ___ is the process of moving heat from one area to another by use of a refrigerant in a closed system.

A. Refrigeration
B. Enthalpy
C. Ventilation
D. Subcooling

______________ **2.** At atmospheric pressure, ___°F is the melting (phase transition) point for water.

A. 0
B. 32
C. 75
D. 120

______________ **3.** A(n) ___ is a heat exchanger through which heat is absorbed from the surroundings into a low-pressure refrigerant liquid.

A. compressor
B. condenser
C. evaporator
D. refrigerator

______________ **4.** A(n) ___ point is a point in a refrigeration system where there is a significant pressure change.

A. saturation
B. phase change
C. evaporation
D. division

Completion

____________________ **1.** A(n) ___ is a chemical that vaporizes (boils) at a low temperature.

____________________ **2.** ___ heat is heat energy that can be measured by a change in temperature.

____________________ **3.** ___ is the removal of sensible heat from a substance after it has condensed.

____________________ **4.** A(n) ___ device is a component that controls the flow rate of refrigerant into an evaporator.

____________________ **5.** The ___ temperature is the temperature at which a refrigerant (at a certain) pressure changes state by vaporizing or condensing.

REFRIGERATION SYSTEM COMPONENTS

SECTION **8.2**

REVIEW QUESTIONS

Name ______________________________ **Date** ____________________

True-False

T F **1.** Refrigerants are identified by numbers that are assigned according to the physical and chemical composition of a refrigerant.

T F **2.** A TXV controls the flow of refrigerant into an evaporator by maintaining a constant subcooling setting.

T F **3.** Slugging is a condition in which liquid refrigerant enters a compressor and causes hammering.

Multiple Choice

______________ **1.** A ___ compressor is a sealed compressor that can be serviced through removable access plates.

A. hermetic
B. reciprocating
C. semihermetic
D. centrifugal

______________ **2.** For optimal system operation, a TXV should maintain a specific superheat, usually between ___.

A. 2°F and 5°F
B. 8°F and 12°F
C. 15°F and 20°F
D. 30°F and 34°F

______________ **3.** ___ is an automatic process for melting the ice that accumulates on an evaporator.

A. Air-cooling
B. Slugging
C. Short cycling
D. Evaporator defrost

______________ **4.** ___ condensers are the most common condensers used in refrigeration systems.

A. Air-cooled
B. Water-cooled
C. Evaporative
D. Expansion

______________ **5.** A(n) ___ is a refrigerant line fitting located before a TXV that contains a small glass window for observing refrigerant flow.
A. capillary tube
B. sight glass
C. accumulator
D. receiver

Completion

______________ **1.** A(n) ___ liquid is a liquid at a certain pressure and temperature that vaporizes if heat is added.

______________ **2.** A(n) ___ compressor is a compressor that contains a pair of screwlike rotors that interlock as they rotate.

______________ **3.** A(n) ___ is a long, thin tube that resists fluid flow, which causes a pressure decrease.

______________ **4.** An automatic ___ valve is a valve that is opened and closed by the pressure in the line ahead of the valve.

______________ **5.** A(n) ___ is a container in which fluid is stored under pressure.

REFRIGERATION CONTROL SYSTEMS

SECTION **8.3**

REVIEW QUESTIONS

Name ________________________________ **Date** ____________________

True-False

T F **1.** In pump-down control, the pressure switch is installed on the low-pressure side and deenergizes the compressor at the cut-out pressure.

T F **2.** Cut-in pressure is the pressure at which a refrigeration compressor is turned off.

T F **3.** Refrigeration systems are designed to maintain a specific temperature rather than a range of temperatures.

Multiple Choice

________________ **1.** The simplest method of control is with a ___.
- A. pump-down control
- B. low-pressure control
- C. head pressure control
- D. thermostat

________________ **2.** The ___ is the amount that an evaporator must be colder than the space it is cooling so that heat flows from the cooled space into the evaporator.
- A. evaporator temperature difference
- B. cut-in temperature
- C. cut-out temperature
- D. differential

________________ **3.** ___ is the process of varying the amount of refrigerant pumped by a compressor.
- A. Short cycling
- B. Unloading
- C. Cutting in
- D. Cutting out

Completion

________________ **1.** ___ is a control system that prevents a refrigeration system's high-pressure side (head) pressure from falling too low.

________________ **2.** A(n) ___ is a switch that activates at temperatures either above or below a certain setpoint.

________________ **3.** A(n) ___ is the size of a desired temperature range for a conditioned space.

SPECIALTY REFRIGERATION SYSTEMS

SECTION **8.4**

REVIEW QUESTIONS

Name ______________________________ **Date** ______________

True-False

T F **1.** Air-to-air heat pumps are most effective in regions that get as cold as 0°F.

T F **2.** Ammonia as a refrigerant is anhydrous ammonia, which means that it does not contain water.

T F **3.** Cooling towers are usually located outside a facility.

T F **4.** Heat pumps are classified as air-to-air, water-to-water, or air-to-water.

T F **5.** Absorption systems are not mechanical compression systems.

Multiple Choice

______________ **1.** A(n) ___ system is a refrigeration system for extremely cold applications that uses two separate mechanical compression refrigeration systems in series.

A. secondary refrigeration
B. cascade
C. chilled water
D. ammonia

______________ **2.** A(n) ___ valve is a four-way valve that reverses the flow of refrigerant in a heat pump.

A. reversing
B. expansion
C. one-way
D. auxiliary

______________ **3.** A vacuum is pressure lower than atmospheric pressure, or ___ psi.

A. 7.2
B. 14.7
C. 25.6
D. 50.1

______________ **4.** A ___ is a small refrigeration system fitted on top of an ammonia receiver to remove unwanted gases.

A. separator
B. compressor
C. purger
D. vacuum

______________ **5.** A ___ is a chiller component that uses evaporation and airflow to cool water.
- A. rupture disc
- B. vacuum
- C. purger
- D. cooling tower

Completion

______________ **1.** A(n) ___ evaporator is an evaporator that is full of liquid refrigerant.

______________ **2.** To achieve lower temperatures, two-stage ammonia systems use two compressors combined in ___.

______________ **3.** A(n) ___ is a liquid capable of absorbing and transporting refrigerant.

______________ **4.** A(n) ___ pump is a mechanical compression refrigeration system that can reverse the flow of refrigerant, switching between heating and cooling modes.

______________ **5.** ___ can move much more heat per pound of liquid vaporized than other refrigerants.

REFRIGERATION SYSTEM MAINTENANCE

SECTION **8.5**

REVIEW QUESTIONS

Name ______________________________ **Date** ____________________

True-False

T F **1.** Most air-cooled evaporator and condenser problems result from dirt buildup on the coils or restricted airflow that reduces heat transfer.

T F **2.** In a water-cooled condenser, cooling water should not leak into the refrigerant.

T F **3.** A zero temperature difference between water entering and leaving a condenser and evaporator on a chiller's operating log indicates low efficiency in the system due to dirty tubes or problems with the circulating pumps.

Multiple Choice

______________ **1.** A temperature difference of less than 10°F in a condenser indicates ___.
- A. proper water flow
- B. excessive air flow
- C. at least one bent fin
- D. a dirty condenser

______________ **2.** Water-to-air heat pump problems are usually related to the ___.
- A. water source
- B. tubes and fins
- C. refrigerant source
- D. filter age

______________ **3.** All cooling towers require water treatment to prevent ___.
- A. rusting
- B. water hammer
- C. algae growth
- D. temperature drops

Completion

______________ **1.** A(n) ___ is a device used to straighten the edges of the thin metal that extend from a condenser or evaporator coil.

______________ **2.** A(n) oil level ___ is a window located on a compressor crankcase that indicates the level of oil in the compressor.

______________ **3.** ___ analysis may be used to monitor the condition of impeller bearings.

REVIEW QUESTIONS

Name ______________________________ **Date** ______________

True-False

T F **1.** A system designed to operate as a cooler can operate as a freezer with some slight modifications.

T F **2.** A service valve is a manually operated two-way valve used to charge or remove refrigerant or monitor system pressure.

T F **3.** During leak testing, refrigerant vapor is added to raise the system pressure to 10 psi.

T F **4.** Charging is adding refrigerant to a refrigeration system.

T F **5.** If the measured low-pressure side temperature is within ±5% of the required temperature, the amount of refrigerant charge needs to be increased.

Multiple Choice

______________ **1.** A ___ is a service valve used on a small refrigeration system that is only opened or closed by fitting a gauge manifold hose onto the port.
- A. gauge manifold
- B. service port
- C. micron gauge
- D. TXV

______________ **2.** ___ is the removal of all air from a refrigeration system before charging it with refrigerant.
- A. Evacuation
- B. Foaming
- C. Purging
- D. Short cycling

______________ **3.** A micron is equal to ___.
- A. one millionth of a meter
- B. one hundredth of a meter
- C. 1000 μm
- D. 25,400 μm

______________ **4.** A ___ is a thermometer that can measure both dry- and wet-bulb temperatures.
- A. hygrometer
- B. manometer
- C. gauge manifold
- D. psychrometer

______________ **5.** When testing for leaks using dye, the leak site can be spotted by ___.
A. looking for bubbles
B. smelling for a strong ammonia-like scent
C. using a fluorescent light
D. running a cloth over the outside of the line

Completion

______________ **1.** ___ is a frequent on/off operation of a compressor.

______________ **2.** ___ is the formation of a foam in a refrigerant/oil mixture due to rapid evaporation of refrigerant.

______________ **3.** A(n) ___ is a set of valves and pressure gauges used to determine refrigerant system pressures and add or remove refrigerant.

______________ **4.** A(n) ___ is an electronic instrument that indicates the depth of vacuum in microns of mercury.

______________ **5.** ___ temperature is a temperature reading that accounts for the amount of water vapor in the air.

REFRIGERANT REGULATIONS

SECTION 8.7

REVIEW QUESTIONS

Name ______________________________ Date ______________

True-False

T F 1. The ammonia in some refrigerants contributes to the depletion of the Earth's ozone layer.

T F 2. A refrigerant technician must pass one or more EPA-approved tests to become certified in refrigerant charging, recovery, and recycling.

Multiple Choice

______________ 1. A(n) ___-approved equipment-testing organization must certify all recovery and recycling equipment.
- A. FDA
- B. EPA
- C. USDA
- D. CAA

______________ 2. Refrigerant ___ is the process of removing dirt, oil, and moisture from lightly contaminated refrigerant that has been removed from a refrigeration system.
- A. purging
- B. recovery
- C. reclaiming
- D. recycling

Completion

______________ 1. Refrigerant ___ is the process of cleaning heavily contaminated refrigerant.

______________ 2. Section 608 of the ___ prohibits technicians from knowingly venting specific refrigerants into the atmosphere while maintaining, servicing, repairing, or disposing of air conditioning or refrigeration equipment.

REFRIGERATION SYSTEMS

Chapter 8

ACTIVITIES

Name ______________________ **Date** ______________

Heat

_______________ **1.** Sensible heat

_______________ **2.** Temperature change

_______________ **3.** Latent heat

_______________ **4.** Change of state

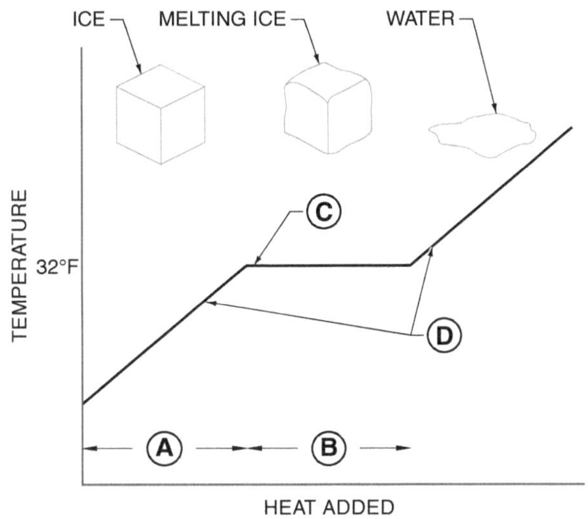

Mechanical Compression Refrigeration System

______ **1.** Evaporator

______ **2.** Compressor

______ **3.** Condenser

______ **4.** Cold air to cooled space

______ **5.** Hot air

______ **6.** Warm air from cooled space

______ **7.** Warm outside air

______ **8.** Vaporizing refrigerant

______ **9.** Condensing refrigerant

______ **10.** Receiver

______ **11.** Filter-drier

______ **12.** Sight glass

______ **13.** Metering device

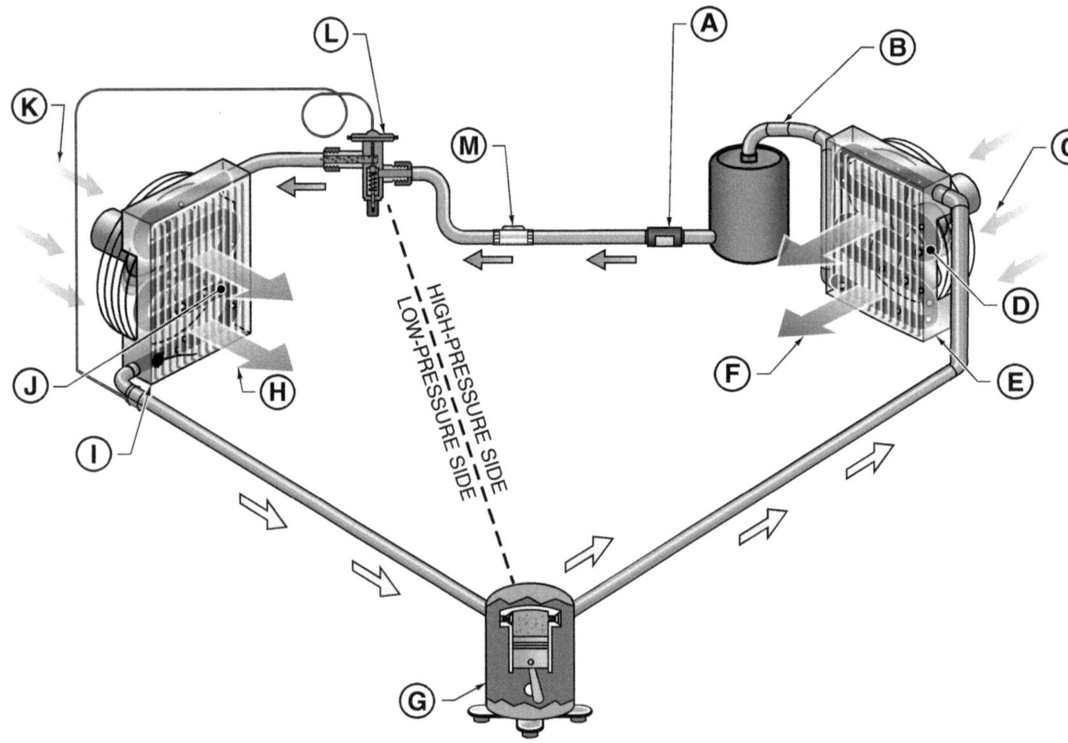

Mechanical Compression Refrigeration Cycle

______________________ **1.** Superheated vapor

______________________ **2.** Condensation

______________________ **3.** Subcooling

______________________ **4.** Subcooled liquid

______________________ **5.** Evaporation

______________________ **6.** Compression

______________________ **7.** Liquid and vapor

______________________ **8.** Expansion

______________________ **9.** Superheat

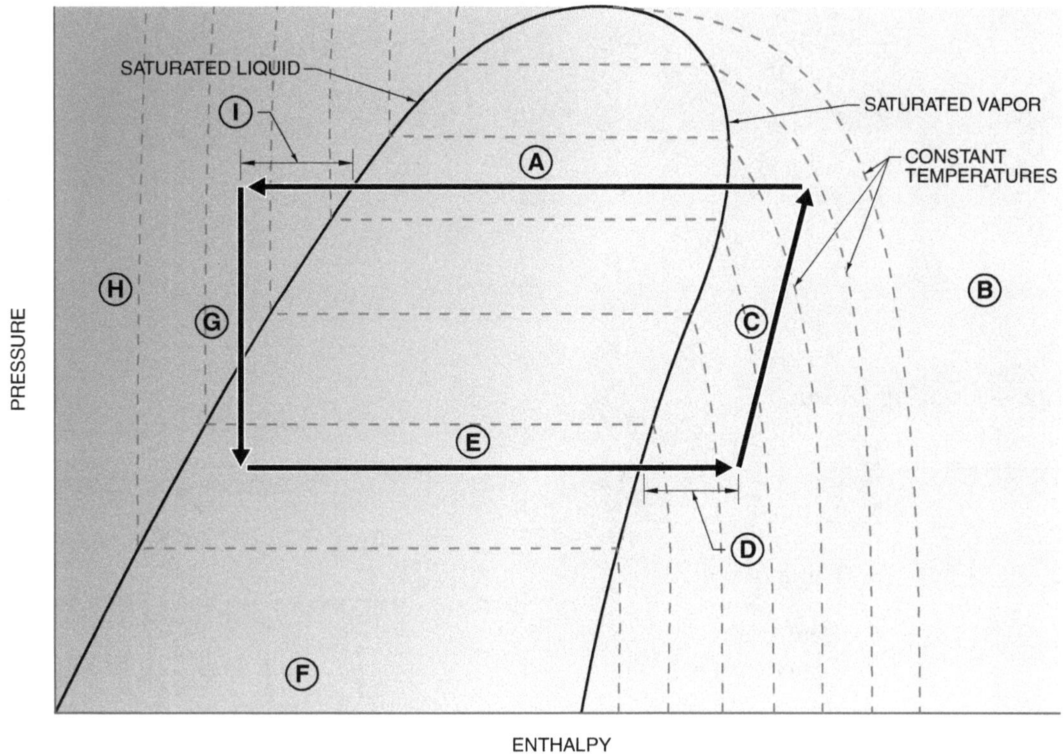

Compressors

________________ **1.** Vane

________________ **2.** Screw

________________ **3.** Reciprocating

________________ **4.** Centrifugal

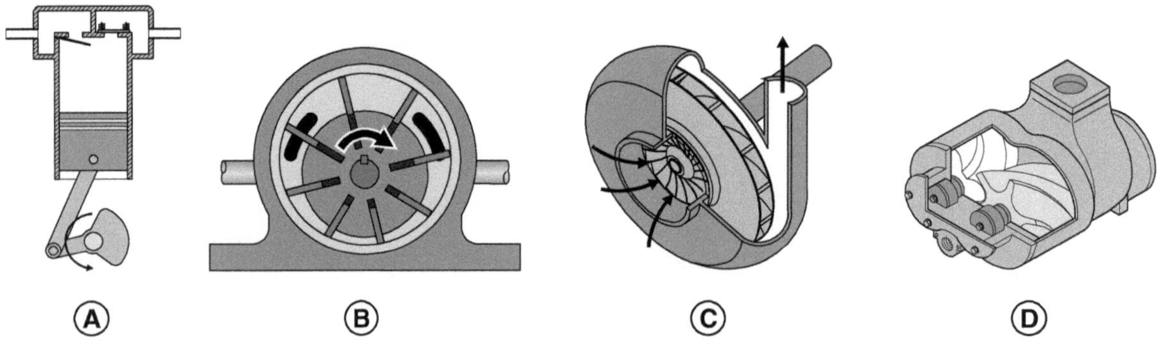

Heat Pumps—Cooling Mode

________________ **1.** Compressor

________________ **2.** Indoor coil

________________ **3.** Outdoor coil

________________ **4.** Heat from heated space

________________ **5.** Heat to outside air

________________ **6.** Refrigerant flow

________________ **7.** Expansion device

________________ **8.** Spool

________________ **9.** Reversing valve

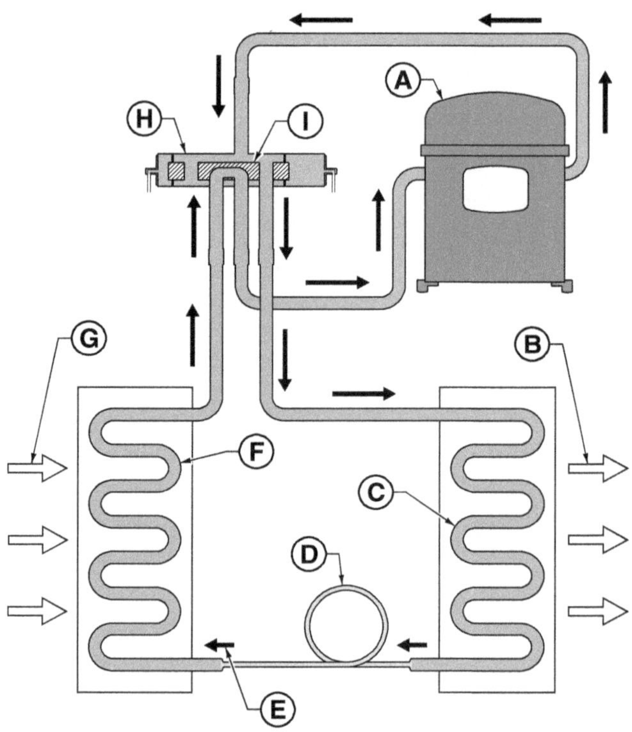

Cooling Towers

____________	**1.** Cooling tower	____________	**9.** Hot refrigerant in
____________	**2.** Condenser	____________	**10.** Condenser tubes
____________	**3.** Tiles	____________	**11.** Water pump
____________	**4.** Cool refrigerant out	____________	**12.** Makeup water inlet
____________	**5.** Cool water in	____________	**13.** Warm condenser water spray
____________	**6.** Cool, dry air in	____________	**14.** Fan
____________	**7.** Hot, moist air out	____________	**15.** Moisture eliminators
____________	**8.** Warm water out		

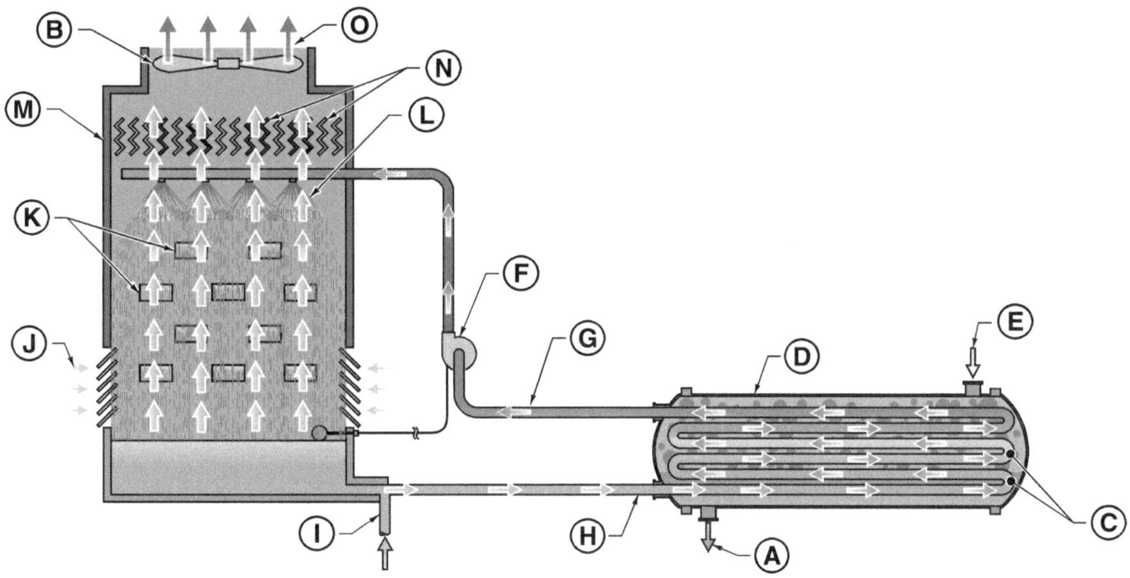

Ammonia Refrigeration Systems

_______________ **1.** Condenser

_______________ **2.** Receiver

_______________ **3.** Evaporator

_______________ **4.** Compressor

_______________ **5.** Roof diffuser

_______________ **6.** Emergency release valve

_______________ **7.** Relief valves

_______________ **8.** Isolation valves

_______________ **9.** Condensing water

_______________ **10.** Purge valves

_______________ **11.** Secondary coolant

_______________ **12.** Oil separator

_______________ **13.** Metering device

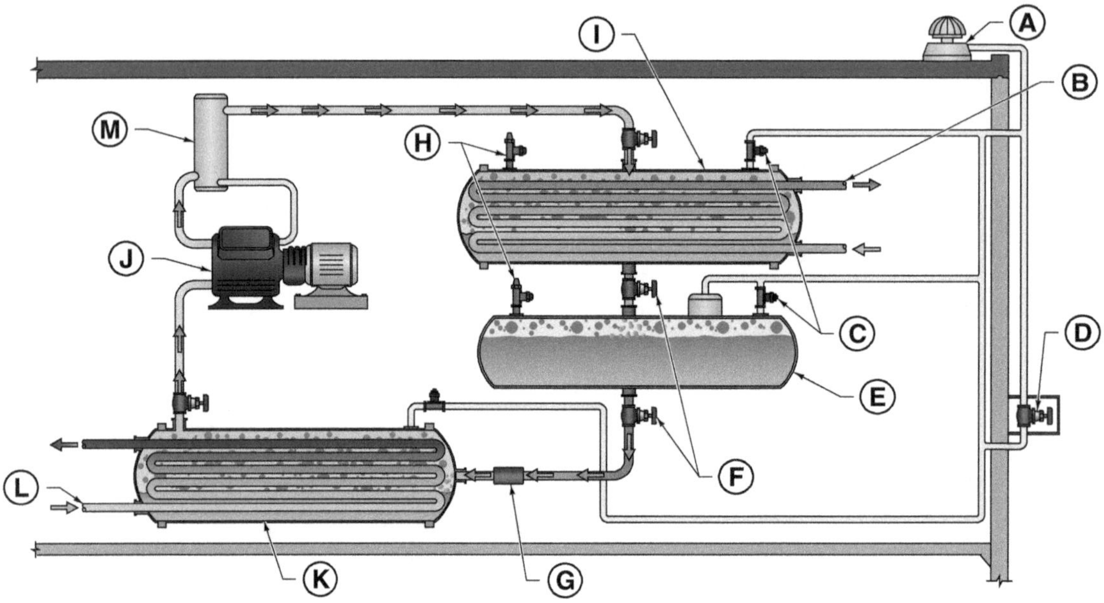

Activity: Refrigerant States

Complete the drawing of the refrigeration system.

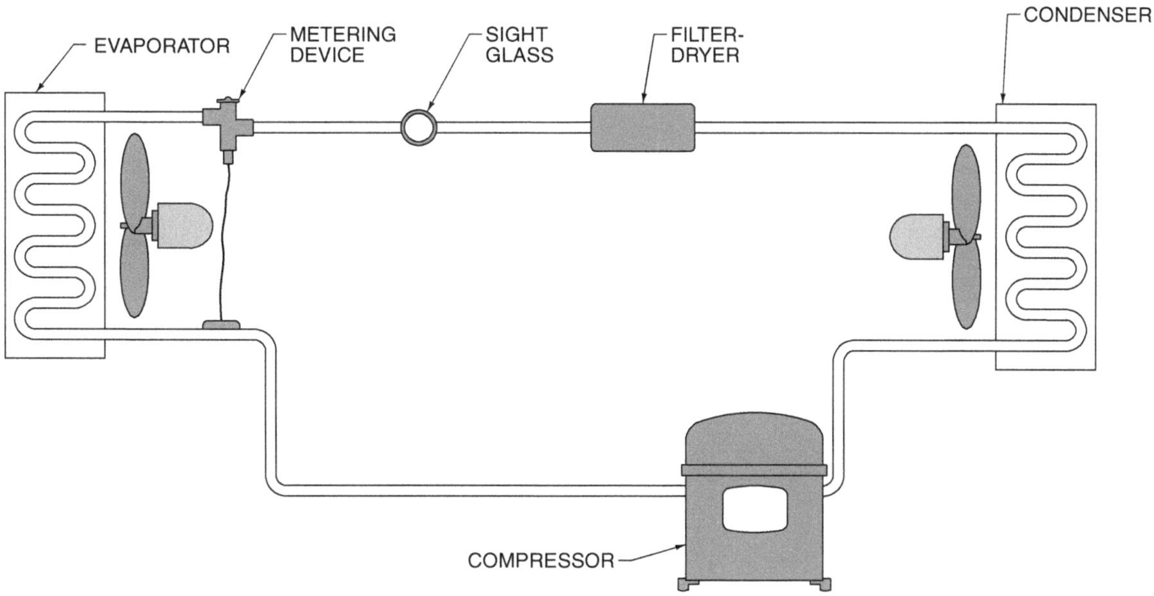

1. Indicate the direction of refrigerant flow with arrows.

2. Sketch the state of the refrigerant as it travels through the system. Use solid shading to indicate liquid refrigerant, light shading with bubbles to indicate vaporizing refrigerant, and no shading to indicate refrigerant vapor.

3. Sketch a line to separate the high-pressure and low-pressure sides.

4. Write in the state of the refrigerant in various parts of the system using the words "vaporizing," "condensing," "subcooled," and "superheated."

5. Indicate the direction of airflow through the evaporator using large arrows and label them with the relative temperature of the air entering and leaving the evaporator. For example, label arrows with "hot air," "warm air," "cool air," or "cold air."

6. Indicate the direction of airflow through the condenser using large arrows and label them with the relative temperature of the air entering and leaving the condenser. For example, label arrows with "hot air," "warm air," "cool air," or "cold air."

7. Why does a refrigeration system need a low-pressure and a high-pressure side to move heat into the evaporator and out of the condenser?

Activity: Temperature-Pressure Charts

Refrigerants have an exact pressure-temperature relationship. At a given pressure, there will be only one saturation temperature, which is the temperature at which the refrigerant will vaporize or condense. This relationship is shown in temperature-pressure charts.

REFRIGERANT VAPOR PRESSURE*			
Temp†	HCFC-22	HCFC-123	HFC-134a
−180	—	—	—
−170	—	—	—
−160	—	—	—
−150	29.4‡	—	—
−140	29.0‡	—	—
−130	28.5‡	—	—
−120	27.7‡	—	—
−110	26.6‡	—	—
−100	25.0‡	—	28.1‡
−90	23.0‡	—	27.1‡
−80	20.2‡	—	25.9‡
−70	16.6‡	—	24.1‡
−60	12.0‡	29.4‡	21.8‡
−50	6.1‡	29.1‡	18.7‡
−40	0.5	28.8‡	14.8‡
−30	4.9	28.3‡	9.9‡
−20	10.2	27.7‡	3.7‡
−10	16.5	26.9‡	1.9
0	24.0	25.8‡	6.5
10	32.8	24.5‡	11.9
20	43.0	22.8‡	18.4
30	54.9	20.7‡	26.1
40	68.4	18.1‡	35.1
50	83.9	15.0‡	45.5
60	101.4	11.2‡	57.5
70	121.2	6.6‡	71.2
80	143.3	1.3‡	86.8
90	168.0	2.5	104.4
100	195.4	6.1	124.3
110	225.8	10.3	146.5
120	259.3	15.1	171.3
130	296.1	20.6	198.9
140	336.5	26.8	229.4
150	380.8	33.9	263.0
160	429.2	41.8	300.1
170	482.0	50.6	340.8
180	539.4	60.4	385.6
190	602.1	71.3	434.7
200	670.3	83.4	488.7
210	—	96.6	548.3

* in psi
† in °F
‡ in inches of mercury vacuum

1. What is the saturation temperature of HCFC-22 at 32.8 psi?

2. What is the saturation temperature of HCFC-22 at 143.3 psi?

3. If 17°F air is blowing over an evaporator coil that contains HCFC-22 at 33 psi, will the refrigerant vaporize?

4. If 80°F air is blowing over a condenser coil that contains HCFC-22 at 226 psi, will the refrigerant condense?

5. What is the saturation pressure of HCFC-22 at 40°F?

6. What is the saturation pressure of HCFC-22 at 140°F?

7. What is the saturation temperature of HCFC-123 at 28.3 in. Hg vacuum?

8. What is the saturation temperature of HCFC-123 at 10.3 psi?

9. If the temperature is 50°F, will the refrigerant in an open container of HCFC-123 vaporize if the container is at sea level?

10. What is the saturation temperature of HFC-134a at 11.9 psi?

11. What is the saturation temperature of HFC-134a at 198.9 psi?

A technician takes temperature and pressure readings on a container of refrigerant recovered from several systems. The pressure is 50 psi over what is expected according to the type of refrigerant and the temperature of the container.

12. What is the most likely cause of the unexpected reading?

13. Should this refrigerant be used in a system?

14. If not, what should be done with this refrigerant?

Activity: Compressors

A reciprocating compressor in a refrigeration system has clogged valves that do not seal properly.

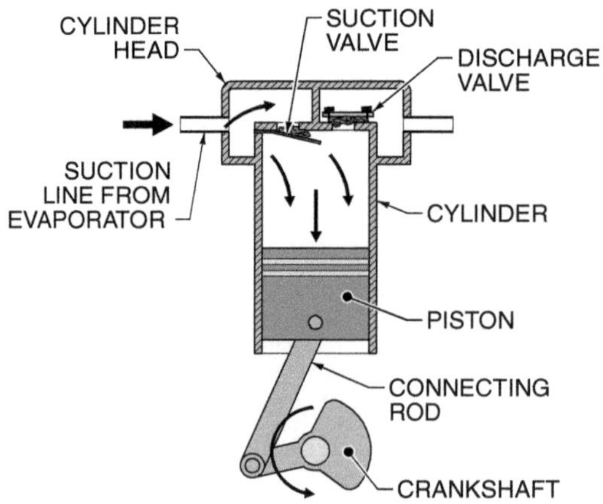

1. What could happen to the low-pressure side and high-pressure side system pressures?

2. What would likely happen to the temperature in the cooled space?

3. How could this problem cause the compressor to overheat?

4. What maintenance do hermetic compressors require?

5. There are several common maintenance activities for large, serviceable compressors. List two examples.

Activity: Condensers

Complete the drawing of the normally operating refrigeration condenser. The temperature inside the condenser should be about 35°F higher than the air blowing through the condenser coils. The outside air temperature is 85°F. The refrigerant is HFC-22.

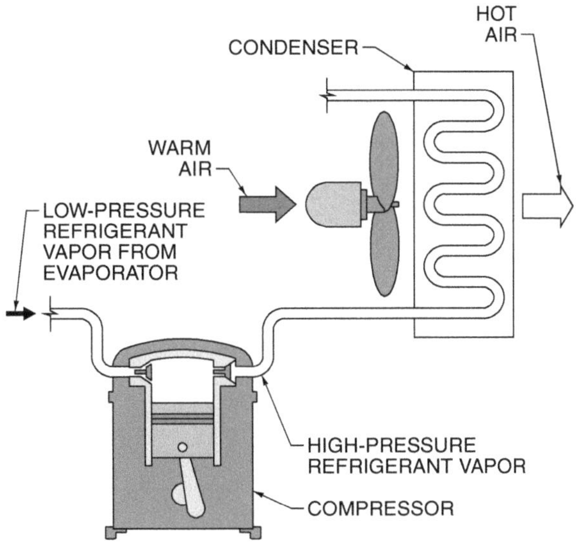

1. Sketch the state of the refrigerant as it travels through the condenser. Use solid shading to indicate liquid refrigerant, light shading with bubbles to indicate vaporizing refrigerant, and no shading to indicate refrigerant vapor.

2. What is the expected pressure inside the condenser?

3. Why should the condenser temperature be so much higher than the surrounding air?

Another condenser under the same operating conditions has coils and fins that are dirty and bent.

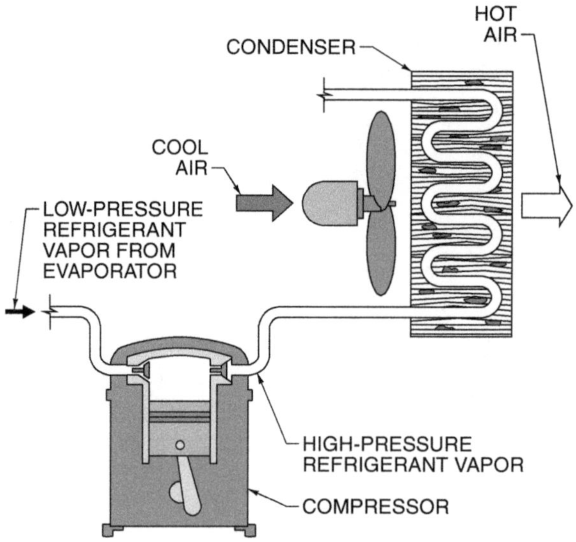

4. Sketch the state of the refrigerant as it travels through the condenser. Use solid shading to indicate liquid refrigerant, light shading with bubbles to indicate vaporizing refrigerant, and no shading to indicate refrigerant vapor.

5. Why would the subcooling in this condenser be low?

6. What problem could occur in an evaporator if the air temperature cooling the condenser is very cold?

7. List two common maintenance activities for air-cooled condensers.

The condenser fan in a separate refrigeration system is not operating.

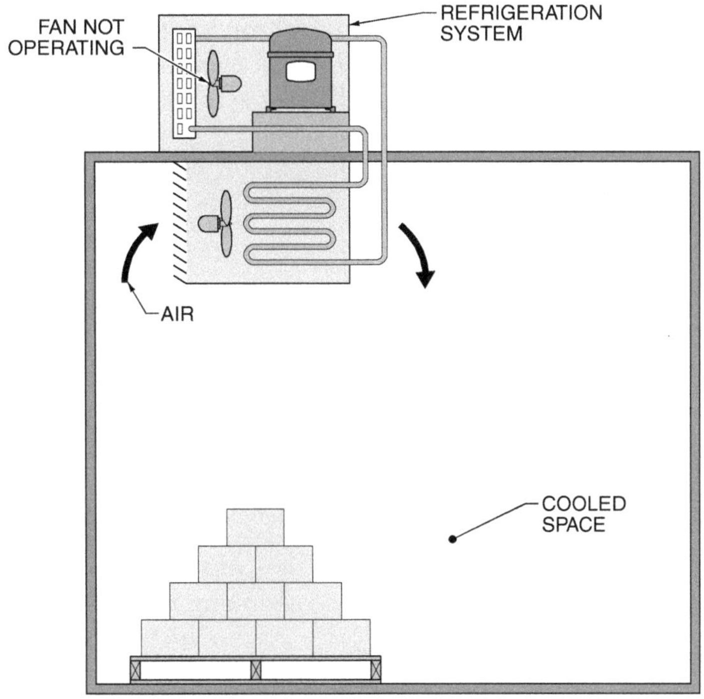

8. What three symptoms could develop from this problem?

9. Could a refrigerant gauge manifold be used to diagnose this problem?

10. What are three possible causes of this problem?

11. Which other component might have been damaged because of this problem?

Activity: Evaporators

Refrigerant HCFC-22 travels through a normally operating evaporator cooling a space to 35°F. The refrigerant enters the evaporator at a temperature of 20°F and is superheated by 10°F. The inside of the evaporator should be at least 10°F cooler than the air flowing across the evaporator coils.

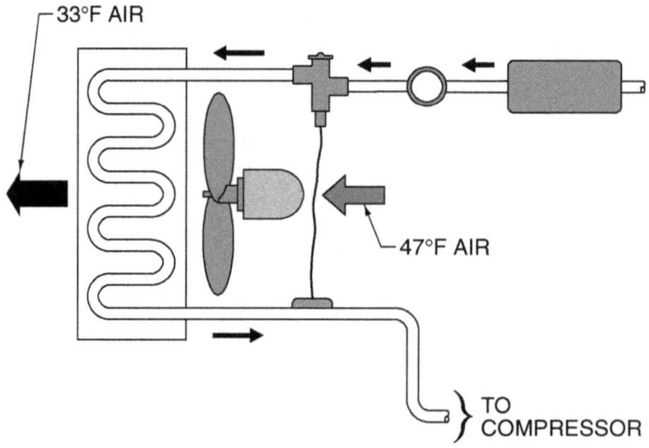

1. Sketch the state of the refrigerant as it travels through the evaporator. Use solid shading to indicate liquid refrigerant, light shading with bubbles to indicate vaporizing refrigerant, and no shading to indicate refrigerant vapor.

2. What is the temperature of the refrigerant at the inlet of the evaporator?

3. What is the expected temperature of the refrigerant at the outlet of the evaporator?

4. Why should the temperature of the evaporator be so much lower than the surrounding air?

5. What is the pressure in this evaporator?

6. List three common maintenance activities that aid in proper airflow through an evaporator.

The refrigerant travels through a starved evaporator. The refrigerant enters the evaporator at a temperature of 20°F and leaves at 40°F.

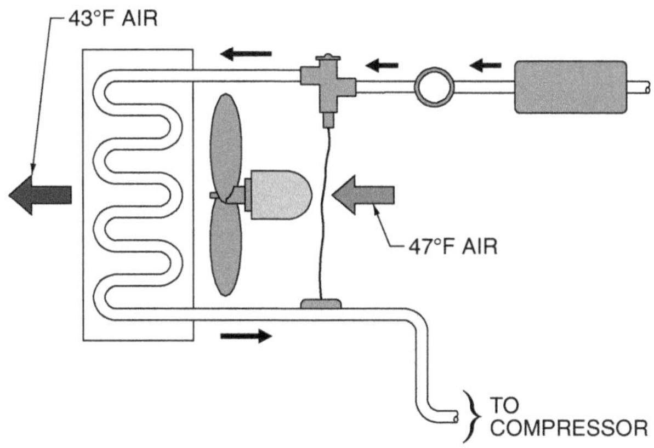

7. Sketch the state of the refrigerant as it travels through the evaporator. Use solid shading to indicate liquid refrigerant, light shading with bubbles to indicate vaporizing refrigerant, and no shading to indicate refrigerant vapor.

8. What three symptoms could develop from this situation?

9. What are three possible causes of this problem?

The refrigerant travels through a flooded evaporator. The refrigerant enters the evaporator at a temperature of 20°F.

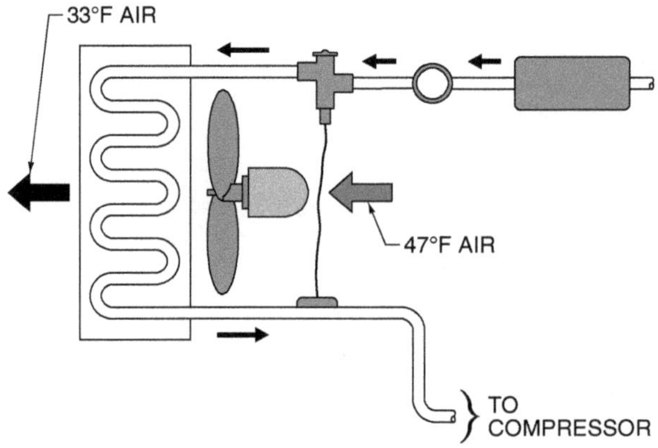

10. Sketch the state of the refrigerant as it travels through the evaporator. Use solid shading to indicate liquid refrigerant, light shading with bubbles to indicate vaporizing refrigerant, and no shading to indicate refrigerant vapor.

11. What three symptoms could develop from this situation?

12. What are two possible causes of this problem?

An evaporator equipped with an automatic defrost cycle is found covered with thick ice.

13. What are three possible causes of this problem?

A cooled space is not at the correct temperature, but the evaporator is very cold.

14. What is the most likely cause of this problem?

15. Why might the evaporator ice up because of this problem?

16. Why might the suction piping leading to the compressor be sweating or frosted?

Activity: Hot-Gas Defrost

A refrigeration system is set up for hot-gas defrost.

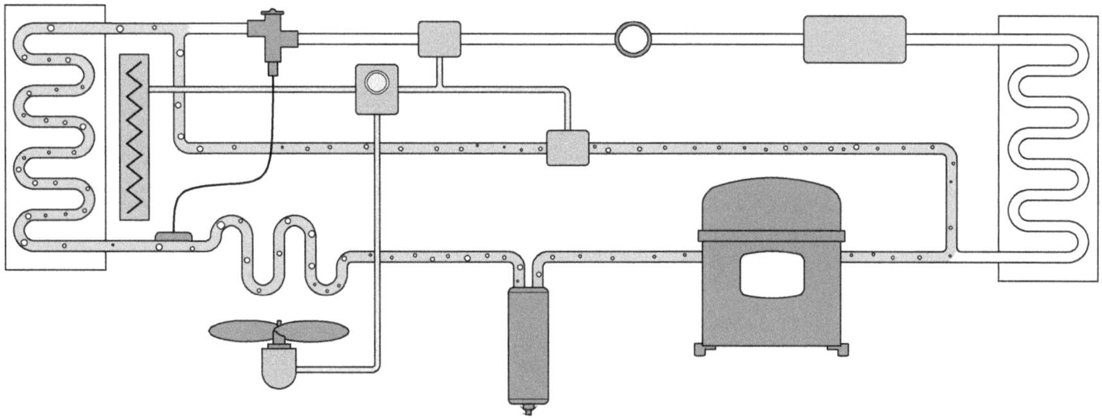

1. Indicate the direction of refrigerant flow with arrows.

2. Which electrical components are energized during hot-gas defrost?

3. If the bypass solenoid valve failed, what would happen when the timer called for the defrost cycle to begin?

4. If the bypass system were operating correctly but not all the frost were melting during defrosting, what is the likely cause of this problem?

5. What is the purpose of the accumulator in the hot-gas defrost system?

6. What are common maintenance activities for a hot-gas defrost system?

Activity: Temperature Control

Boxes in a cooled space are placed directly on the floor and stacked up to the bottom of the refrigeration unit.

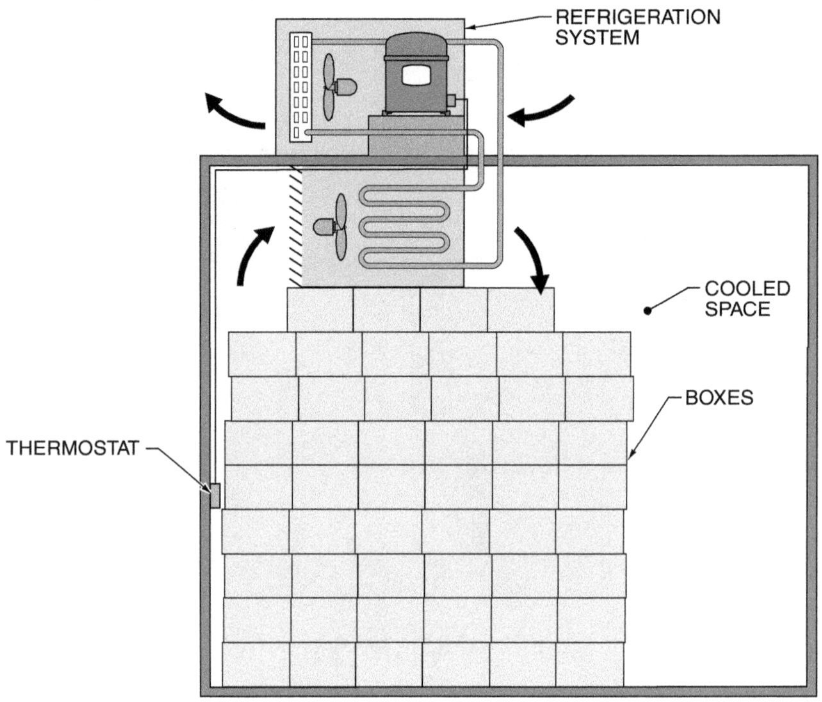

1. What three symptoms could develop from this situation?

2. Why should the boxes have been stacked on pallets instead?

3. If the cooled-space doors have been open for a long time, why should pressure readings be taken only after a few hours have passed?

Activity: Heat Pumps

A heat pump system is set up for cooling mode.

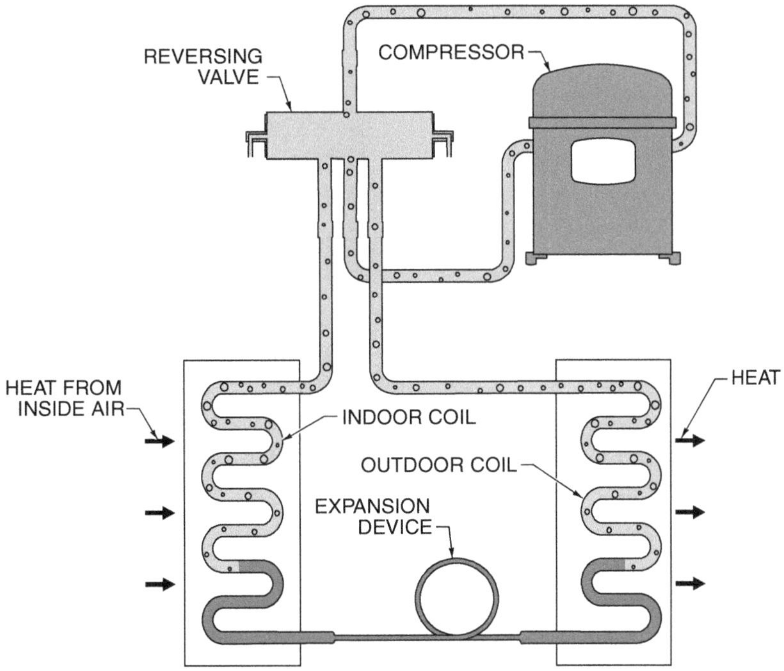

1. Sketch the spool position in the reversing valve.

2. Indicate with arrows the direction of refrigerant flow and airflow through the indoor and outdoor coils.

3. Label each coil with its current function as a "condenser" or an "evaporator."

A heat pump system is set up for heating mode.

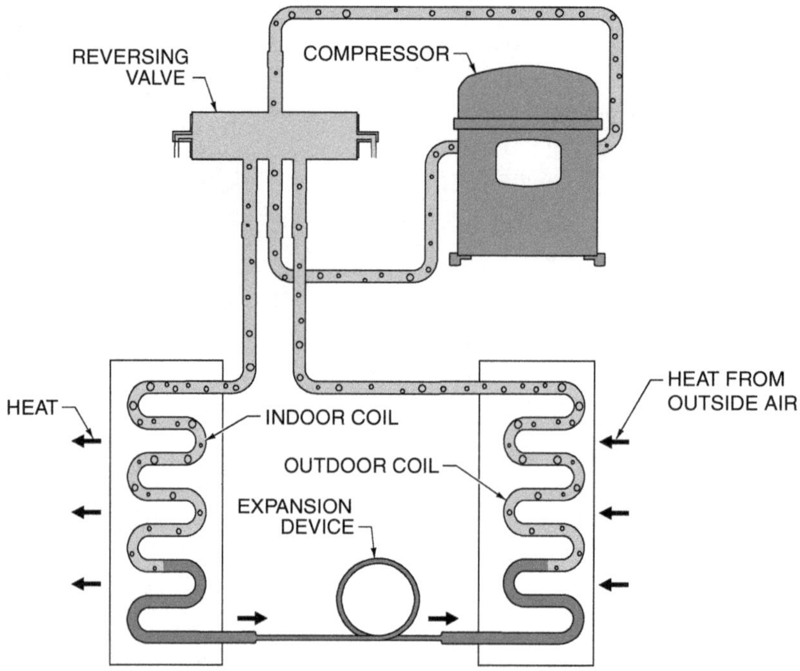

4. Sketch the spool position in the reversing valve.

5. Indicate with arrows the direction of refrigerant flow and airflow through the indoor and outdoor coils.

6. Label each coil with its current function as a "condenser" or an "evaporator."

7. Failure of which component could cause the heat pump to inadvertently switch from cooling to heating?

Activity: Chilled Water Systems

A low-pressure chilled water system is operating in a large building.

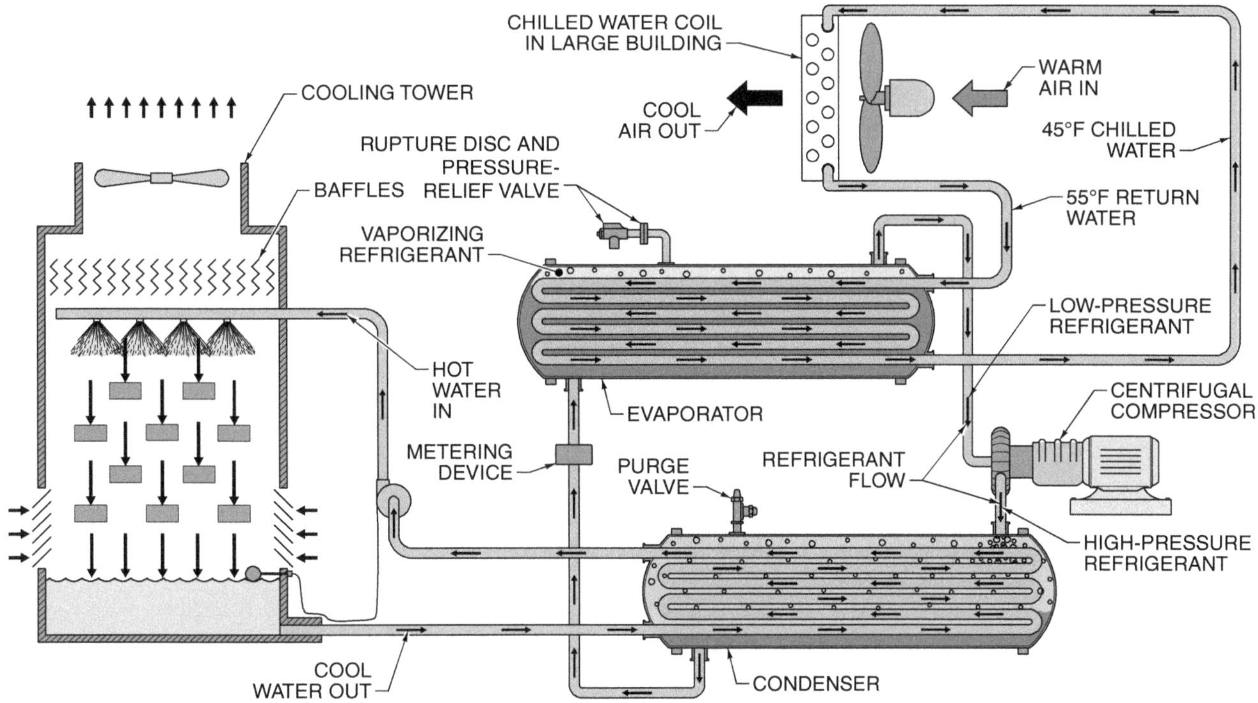

1. Do evaporators in a low-pressure chiller operate above or below atmospheric pressure?

2. What is a common pressure in the condenser?

3. What is a common setting for the rupture disc?

4. Which device prevents all the refrigerant from escaping the system if the rupture disc bursts?

5. What would happen to the system if the circulating pump that sends water to the cooling tower fails?

Activity: Cooling Towers

A cooling tower is operating as part of a refrigeration system.

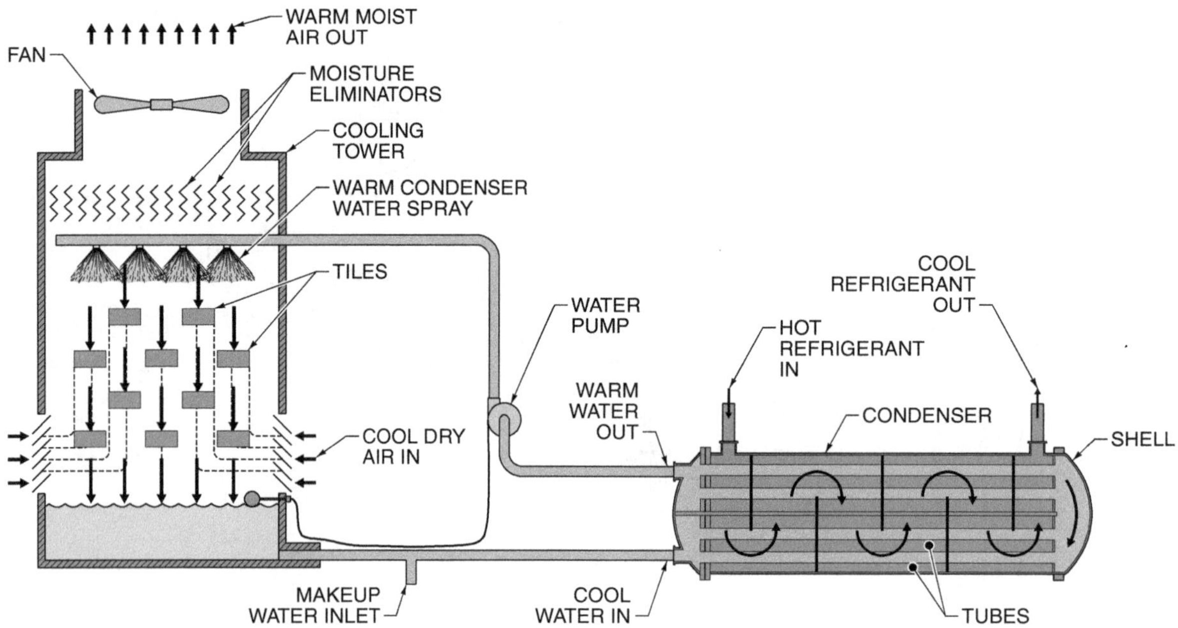

1. If the water flow to the water-cooled condenser were restricted, what would happen to the ability of the system to cool the cooled space?

2. What are three possible causes of restricted water flow?

3. What would happen to the temperature of the condenser if water flow were restricted?

4. What would happen to the temperature difference between the water entering and leaving the condenser if water flow were restricted?

5. List three common maintenance activities for water-cooled condensers.

6. List three common maintenance activities for cooling towers.

Activity: Taking Pressure Readings

A gauge manifold is set up to take pressure readings on a refrigeration system

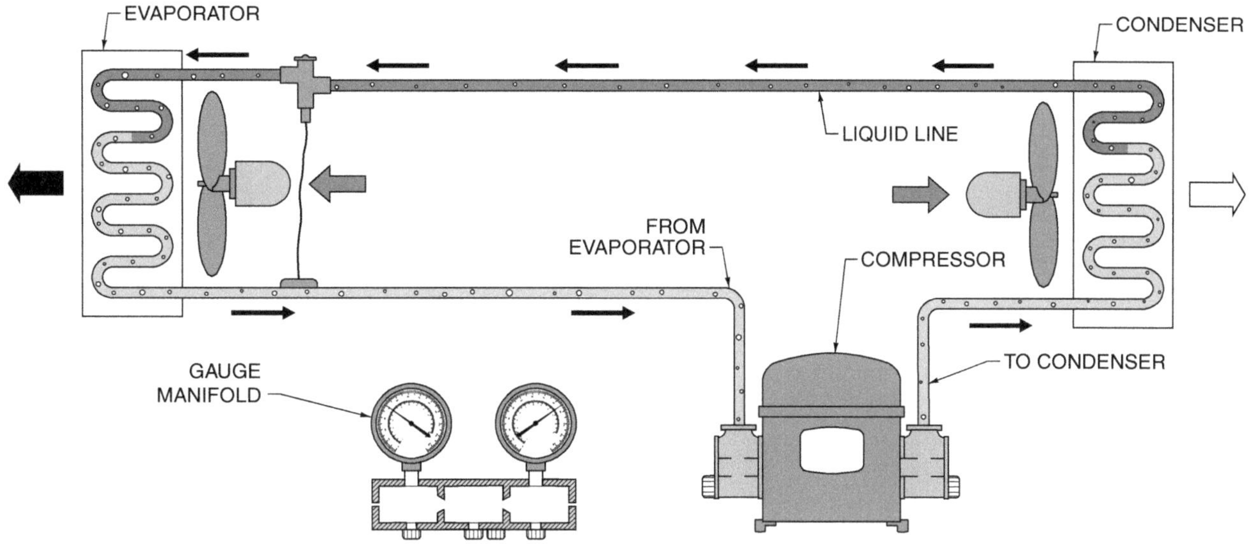

1. Sketch and label the valve positions on the service manifold and on the compressor service valves.

2. Sketch the hose connections.

3. How is liquid refrigerant removed from the high-pressure side hose after taking system pressures?

4. What are two possible causes of a small pressure difference between the high-pressure and low-pressure sides?

5. There are several possible causes of an extremely high compressor discharge-side pressure and extremely low suction-side pressure. List two causes.

6. What pressures would be expected between the high-pressure and low-pressure side in a system that had not been operating for a long time?

Activity: Evacuating Systems

A gauge manifold is set up to evacuate a refrigeration system.

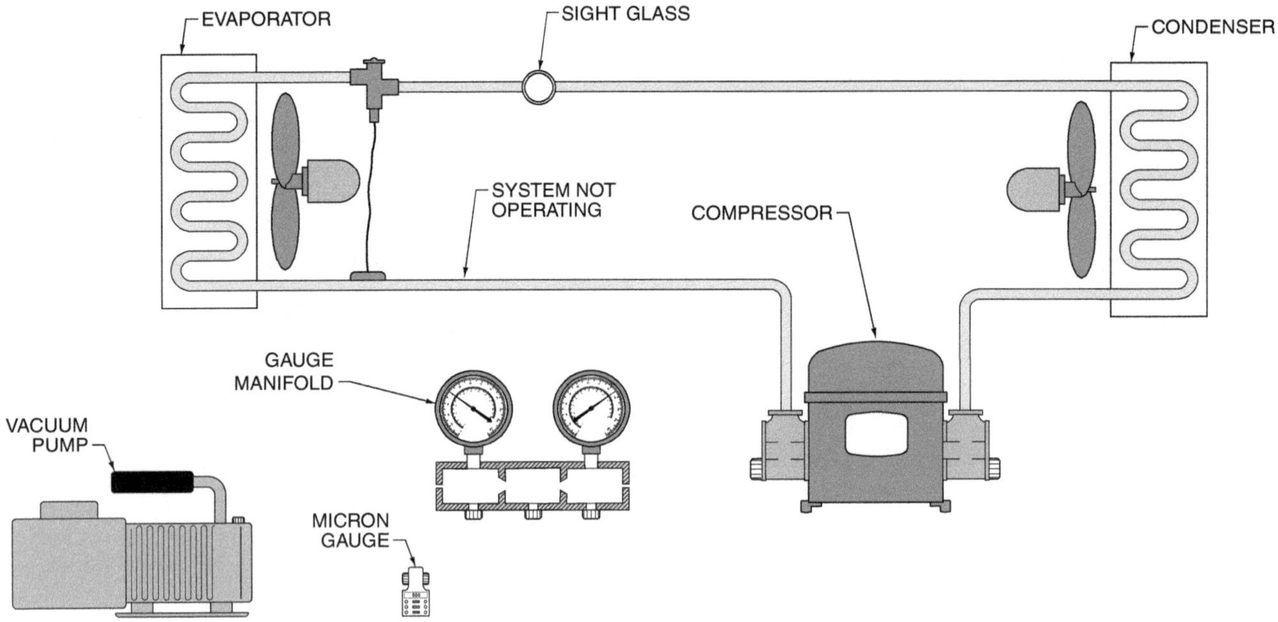

1. Sketch and label the valve positions on the service manifold and on the compressor service valves.

2. Sketch the hose connections between all the evacuation equipment.

3. What pressure level indicates that the system is evacuated?

4. What happens to the micron gauge reading immediately after the vacuum pump is turned off?

5. If the system is leaking, what will happen to the reading on the micron gauge?

6. If the system still has water in it, what will happen to the reading on the micron gauge?

Activity: Liquid Charging Systems

A gauge manifold is set up to charge a refrigeration system with liquid.

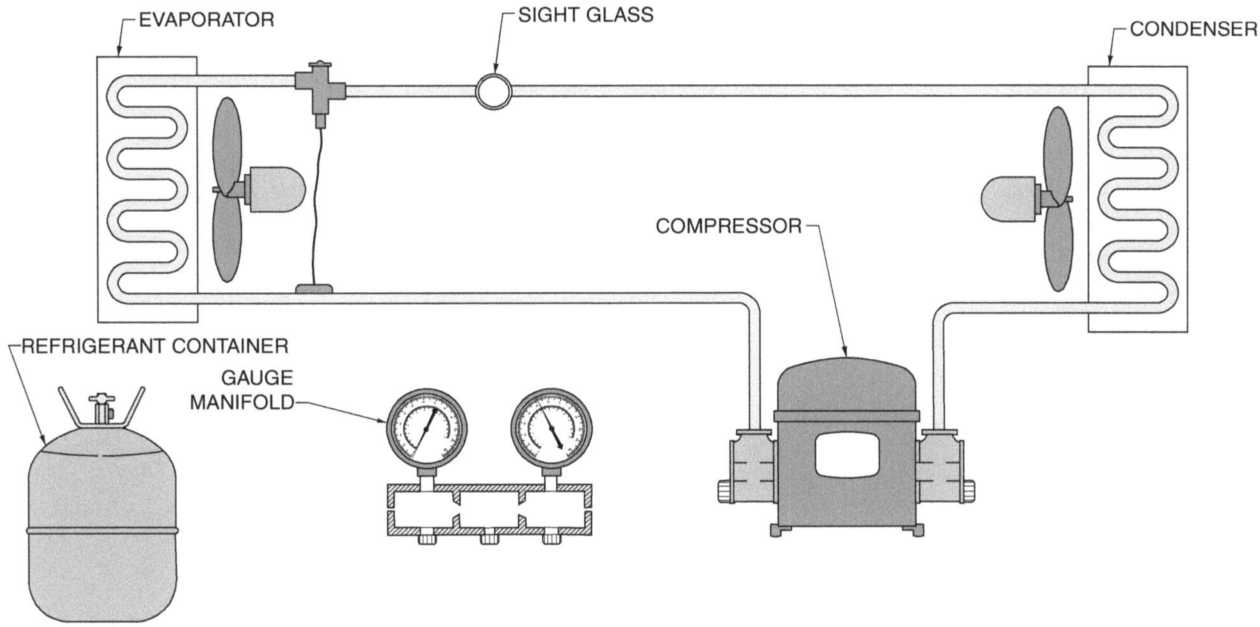

1. Sketch and label the valve positions on the service manifold and on the compressor service valves.

2. Sketch the hose connections between all the charging equipment.

3. Indicate the direction of refrigerant flow with arrows.

4. Should the container of refrigerant be right side up or upside down on the electronic scale?

5. When would liquid refrigerant stop flowing into the system?

Activity: Vapor Charging Systems

A gauge manifold is set up to charge a refrigeration system with vapor.

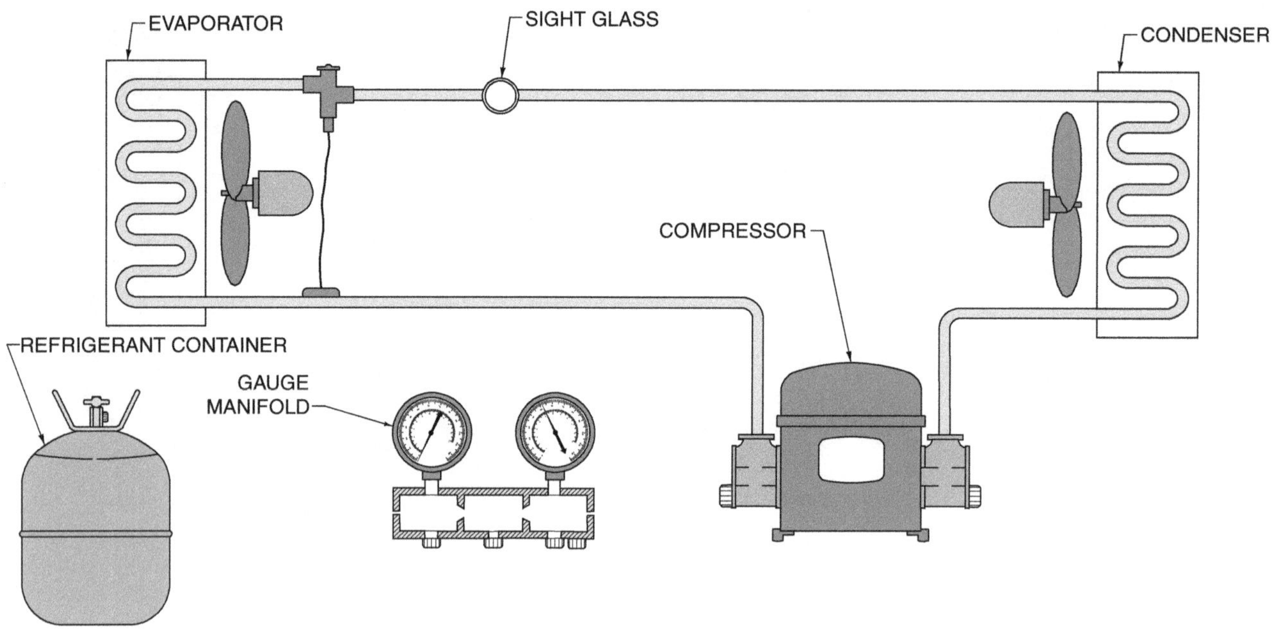

1. Sketch and label the valve positions on the service manifold and on the compressor service valves.

2. Sketch the hose connections between all the charging equipment.

3. Indicate the direction of refrigerant flow with arrows.

4. Why must the system be running while vapor charging?

5. What four conditions should be monitored to determine if the system is fully charged?

Activity: Leaks

A refrigeration system has been leaking and is low on refrigerant.

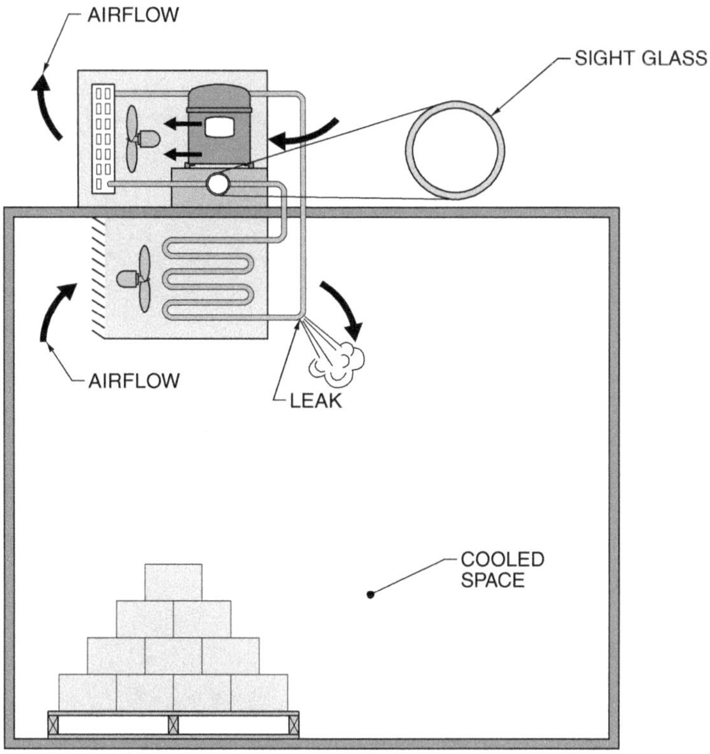

1. Sketch what will be seen in the sight glass.

2. Several symptoms could develop from this situation. List three symptoms.

3. What are three possible causes of leaks in refrigeration systems?

4. How is a standing pressure test conducted on the system?

5. What is the technique for finding a leak with an electronic leak detector?

6. How is the general location of the leak found?

7. How is the exact location of the leak found?

8. What visual sign might be found at the site of the leak?

Refrigeration Troubleshooting

The evaporator fan of a refrigeration unit is not operating.

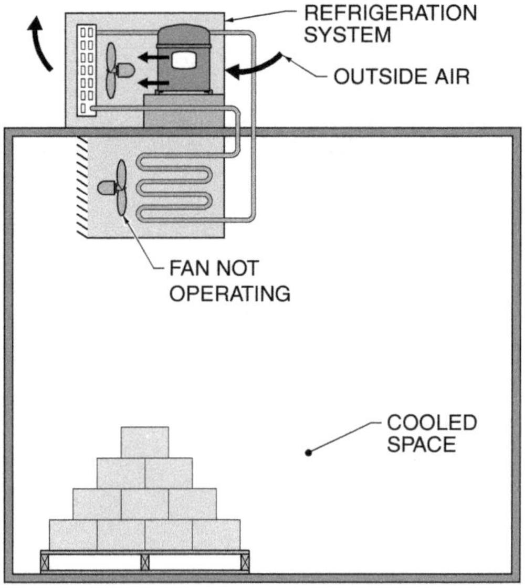

1. Would the pressure on the high-pressure side be higher or lower than normal?

2. Would the pressure on the low-pressure side be higher or lower than normal?

3. Would the superheat be higher or lower than normal?

4. Would the subcooling be higher or lower than normal?

5. What would happen to the temperature in the cooled space?

6. Would it be necessary to first attach a manifold gauge to determine the cause of the problem?

7. How might the compressor be damaged by this situation?

8. Could the evaporator ice up in this situation?

The condenser coils of a refrigeration unit are clogged with dirt.

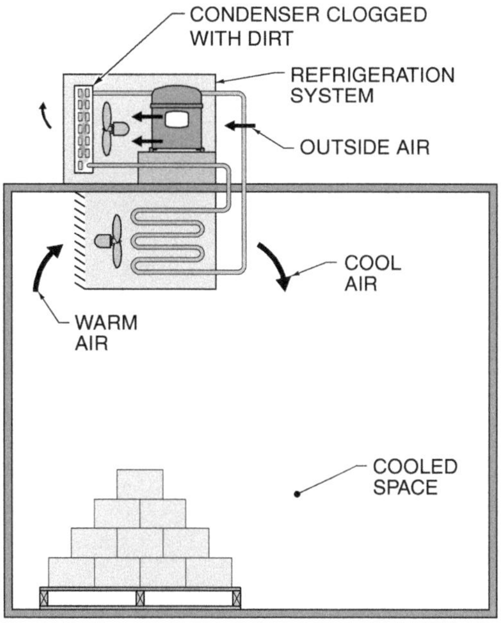

9. Would the pressure on the high-pressure side be higher or lower than normal?

10. Would the pressure on the low-pressure side be higher or lower than normal?

11. Would the superheat be higher or lower than normal?

12. Would the subcooling be higher or lower than normal?

13. What would happen to the temperature in the cooled space?

14. Would it be necessary to first attach a manifold gauge to determine the cause of the problem?

15. How might the compressor be damaged by this situation?

A refrigerant leak is suspected in the lines between the compressor and the condenser of a refrigeration unit.

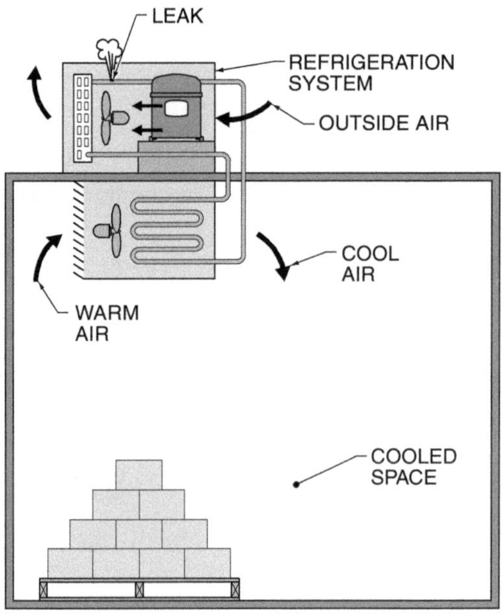

16. Would the pressure on the high-pressure side be higher or lower than normal?

17. Would the pressure on the low-pressure side be higher or lower than normal?

18. Would the superheat be higher or lower than normal?

19. Would the subcooling be higher or lower than normal?

20. What would happen to the temperature in the cooled space?

21. Would it be necessary to attach a manifold gauge to determine the cause of the problem?

22. How might the compressor be damaged by this situation?

23. How would the location of the leak be found?

24. If the leak repair required opening the system, what general procedures would be followed?

25. If the system contained R-22, what additional concern would affect the decision to repair or replace this system?

BOILER SYSTEM OVERVIEW

SECTION 9.1

REVIEW QUESTIONS

Name ______________________________ **Date** ____________________

True-False

T F **1.** A breeching is the duct connecting a boiler to a chimney.

T F **2.** Electric boilers operate at up to 1600 V.

T F **3.** A firebox boiler is a firetube boiler suspended over a furnace.

T F **4.** A package boiler is preassembled at the factory.

T F **5.** Cast iron boilers are limited to low-pressure steam or hot water for heating and process applications.

Multiple Choice

______________ **1.** A low-pressure boiler is a boiler that has an MAWP of up to ___ psi.

A. 15
B. 30
C. 45
D. 60

______________ **2.** A ___ is the power available from the evaporation of 34.5 lb of water per hour at a feedwater temperature of 212°F.

A. British thermal unit (Btu)
B. boiler horsepower (BHP)
C. boiler power unit (BPU)
D. therm

______________ **3.** A therm is a unit of heat energy equivalent to ___ Btu.

A. 1000
B. 10,000
C. 100,000
D. 1,000,000

______________ **4.** ___ heat is heat energy that causes a change of state but not change in temperature.

A Potent
B. Latent
C. Patent
D. Sensible

______________ **5.** A ___ boiler is a boiler that is constructed on-site.
A. package
B. field-erected
C. prefabricated
D. scotch marine

Completion

______________ **1.** A(n) ___ is the lowest part of the water side of a watertube boiler and collects sludge or mud.

______________ **2.** ___ is water that is supplied to a boiler at the proper temperature and pressure.

______________ **3.** A(n) ___ system is a system that regulates the flow of air to and from a boiler.

______________ **4.** A(n) ___ boiler is a boiler in which the hot gases of combustion pass through tubes surrounded by water.

______________ **5.** Thermal ___ is the ratio of heat absorbed to the heat available.

______________ **6.** A(n) ___ is the recommended maximum pressure at which a boiler can safely operate.

______________ **7.** ___ heat is heat energy that can be measured by a change in temperature.

______________ **8.** A(n) ___ steam line is a line that connects a boiler to the steam header.

BOILER FITTINGS AND ACCESSORIES

SECTION 9.2

REVIEW QUESTIONS

Name ______________________________ **Date** ____________________

True-False

T F **1.** A boiler fitting is not directly attached to a boiler but is necessary for its operation.

T F **2.** A globe valve is a valve that controls flow by raising or lowering a circular disc.

T F **3.** The ASME Code specifies the design, materials, and construction of safety valves.

T F **4.** Open feedwater heaters achieve higher water temperatures than closed feedwater heaters.

T F **5.** An OS&Y valve is open when the stem is up.

Multiple Choice

________________ **1.** The ASME symbol stamp for a safety valve is the letter ___ and is included on the data plate of the safety valve.

A. A
B. E
C. S
D. V

________________ **2.** ___ is the inclusion of small water droplets in steam lines, which can cause water hammer.

A. Slugging
B. Carryover
C. Makeup
D. Reclaiming

________________ **3.** Water columns are located at the ___ with the lowest part of the water column positioned at least 3″ above the heating surface.

A. BTU
B. MAWP
C. NOWL
D. MAWL

_______________ **4.** A(n) ___ is a valve mounted on a water column that can be used to determine boiler water level.
A. try cock
B. isolation valve
C. check valve
D. bottom blowdown valve

_______________ **5.** A(n) ___ is used to release water from the bottom of a boiler in order to reduce the water level, remove sludge and sediment, reduce chemical concentrations, or drain the boiler.
A. try cock
B. gate valve
C. isolation valve
D. bottom blowdown valve

_______________ **6.** A ___ steam trap is a steam trap that has a single movable disc that rises to allow the discharge of air and cool condensate.
A. thermodynamic
B. float thermostatic
C. thermostatic
D. inverted bucket

Completion

_______________ **1.** A(n) ___ valve is a valve that allows flow in only one direction.

_______________ **2.** ___ water is water that is used to replace boiler water lost from leaks or blowdowns.

_______________ **3.** Boilers having over ___ sq ft of boiler heating surface must have at least two means of supplying feedwater to the boiler.

_______________ **4.** A(n) ___ tube is a curved, hollow tube closed on one end that straightens when pressure is applied in the tube.

_______________ **5.** A(n) ___ valve is a fitting that prevents a boiler from exceeding its MAWP.

_______________ **6.** Water ___ is a banging caused by rapid water movement in steam lines.

COMBUSTION AND DRAFT

SECTION **9.3**

REVIEW QUESTIONS

Name ______________________________ **Date** ____________________

True-False

T	F	**1.** Perfect combustion is achievable only in a laboratory setting and cannot be achieved in a boiler.
T	F	**2.** Coal is a mixture of semisolids, liquids, and gases formed from the remains of organic matter that has been changed by pressure and heat over millions of years.
T	F	**3.** Hand firing is rarely used for feeding boilers today because of the loss of heat when the firing door is open.
T	F	**4.** A positioning control system is used in plants that have a steam demand large enough to require the burners to be fired continuously.
T	F	**5.** Natural draft provides greater control of the combustion process and is more widely used than mechanical draft.

Multiple Choice

______________ **1.** A(n) ___ valve is located between the pump and the discharge valve.

A. bottom blowdown
B. isolation
C. relief
D. safety

______________ **2.** A ___-feed stoker is a stoker that pushes feeder blocks to advance coal into a furnace.

A. ram
B. chain
C. screw
D. hand

______________ **3.** A ___ is a combustion controller that regulates the sequence of events for lighting and shutting down a burner.

A. CPU
B. human-machine interface
C. manometer
D. programmer

______________ **4.** ___ air is air in the combustion process that regulates the rate of combustion.
- A. Primary
- B. Secondary
- C. Incomplete
- D. Excess

______________ **5.** ___ draft is mechanical draft from air pulled through the boiler furnace with fans located in the breeching.
- A. Forced
- B. Balanced
- C. Induced
- D. Natural

Completion

______________ **1.** A(n) ___ is a mechanical device used to feed coal pieces to a furnace.

______________ **2.** A(n) ___ is a device that uses a liquid-filled tube to measure the difference in pressure between two locations.

______________ **3.** A flame ___ is a safety device that senses if the pilot light and main flame are lit.

______________ **4.** ___ combustion is combustion of all fuel using only the theoretical minimum amount of air.

______________ **5.** ___ is the difference in pressure between two points that causes air or gas to flow.

BOILER OPERATION PROCEDURES

SECTION 9.4

REVIEW QUESTIONS

Name ______________________________ **Date** ____________________

True-False

T F **1.** A boiler room log is a record of the temperatures, pressures, and fuel consumption of a boiler and accessories in a plant over a 24 hr period, along with any extraordinary events.

T F **2.** The NOWL is located at the top of the gauge glass.

T F **3.** Cold-plant start-up is considered less hazardous than live-plant start-up.

Multiple Choice

______________ **1.** Water column and gauge glass ___ removes any sediment or debris that could enter the gauge glass and cause inaccurate water levels.

A. blowdown
B. replacement
C. lay-up
D. blowout

______________ **2.** Before inspection, a boiler must be ___.

A. in service for at least one month
B. blown down
C. operating normally
D. taken off-line

______________ **3.** ___ lay-up is storage of a boiler with all water drained.

A. Wet
B. Dry
C. Storage
D. Cold

Completion

______________ **1.** ___ is accomplished by opening the low water fuel cutoff blowdown valve while the burner is firing to simulate a low water level condition.

______________ **2.** ___-plant start-up procedures are performed on boilers that have been out of service for more than two days.

______________ **3.** Boiler ___ is the preparation of a boiler for out-of-service status for an extended time.

BOILER SYSTEM PRINTREADING

SECTION **9.5**

REVIEW QUESTIONS

Name ______________________________ **Date** ______________

True-False

T F **1.** Systems spanning multiple print sheets include numbered continuation arrows to connect the same lines from sheet to sheet.

T F **2.** The Environmental Protection Agency (EPA) requires that piping be well-marked for easy identification.

T F **3.** If a fluid could fall under more than one category, the most hazardous property should be used to determine the marking color code used.

Multiple Choice

______________ **1.** A piping ___ is a drawing that shows the physical location of piping in relation to a structure.

A. floor plan
B. schematic
C. diagram
D. and instrumentation diagram

______________ **2.** A piping ___ contains the fluid flow information of a piping diagram, but it also includes information on the electronic sensors, controllers, and actuators that are becoming increasingly common in these systems.

A. floor plan
B. schematic
C. diagram
D. and instrumentation diagram

______________ **3.** ___ are important for being able to match components identified in boiler system diagrams with the physical devices in the field.

A. Pipe markings
B. Boiler logs
C. Valve tags
D. Piping and instrumentation diagrams

Completion

______________ **1.** A(n) ___ is a drawing that shows the operational arrangement of piping, fittings, and major components of a system conveying fluids.

______________ **2.** ANSI/ASME A13.1-2015, *Scheme for the Identification of Piping Systems,* assigns colors to be used based on the hazards of the ___.

______________ **3.** A(n) ___ is a durable tag for valves and other components that includes identifying abbreviations and/or numbers.

BOILER SYSTEM MAINTENANCE

SECTION **9.6**

REVIEW QUESTIONS

Name ______________________________ **Date** ________________

True-False

T F **1.** Fuel oil strainers used with No. 6 fuel oil are commonly serviced every 24 hr.

T F **2.** Steam trap operation is tested by sight, temperature, and sound.

T F **3.** The pH scale is a scale from 0 to 10 used to indicate the acidity or alkalinity of a solution.

T F **4.** Scale acts like a conducting material and increases heat transfer efficiency.

T F **5.** A gauge glass is normally attached to a water column by a screw or flanged fitting.

Multiple Choice

________________ **1.** A boiler should have approximately ___% of the safety valve popping pressure.
- A. 35
- B. 60
- C. 75
- D. 80

________________ **2.** A ___ verifies a flame in a burner.
- A. flame scanner
- B. pilot light
- C. safety valve
- D. steam trap

________________ **3.** ___ is the accumulation of high alkaline elements that cause boiler metal corrosion.
- A. Scale
- B. Caustic embrittlement
- C. Foaming
- D. Carryover

________________ **4.** Boiler water pH should be between ___.
- A. 5 and 7
- B. 7 and 10
- C. 10 and 11.5
- D. 11 and 13

______________ **5.** Carryover and foaming of boiler water are prevented by controlling the ___.
A. boiler pressure
B. number of leaks
C. water temperature
D. surface tension

Completion

______________ **1.** Safety valves on high-pressure boilers should be manually tested once a(n) ___ and pressure tested once a year.

______________ **2.** ___ blowdown is the purging of boiler water at the surface to remove floating impurities and foreign matter.

______________ **3.** ___ is the treatment of boiler water after it has entered a boiler system and may be circulating through the steam and condensate lines.

______________ **4.** A solution with a pH less than 7 is ___.

______________ **5.** A(n) ___ is a member of a category of microporous, aluminosilicate minerals that are used to soften water.

BOILER SYSTEM TROUBLESHOOTING

SECTION 9.7

REVIEW QUESTIONS

Name ______________________________ **Date** ____________________

True-False

T F **1.** A flame scanner is used for finding areas of damaged or missing insulation.

T F **2.** It is critical that a furnace is always purged of any accumulated fuel before attempting restart.

T F **3.** Water discharging from the top try cock confirms a low water level in the boiler.

Multiple Choice

__________ **1.** ___ steam is steam that leaves the boiler directly at high pressure.
- A. Overpressure
- B. Controlled
- C. Hot
- D. Live

__________ **2.** ___ must be used to confirm water level if water cannot be seen in the gauge glass.
- A. Safety valves
- B. Try cocks
- C. Bottom blowdown valves
- D. Feedwater pumps

__________ **3.** A high water condition occurs when there is an excessive amount of ___ in a boiler.
- A. pressure
- B. gas
- C. water
- D. steam

Completion

__________ **1.** ___ is a condition that occurs when a boiler is operating at or above its MAWP.

__________ **2.** A feedwater pump becomes ___ when pumped water gets too hot and turns to steam.

__________ **3.** A low water condition occurs when the water in a boiler is below the ___.

BOILER SYSTEMS

Chapter 9

ACTIVITIES

Name ______________________________ Date ______________________

Boiler Systems

______________ 1. Draft system

______________ 2. Fuel system

______________ 3. Steam system

______________ 4. Feedwater system

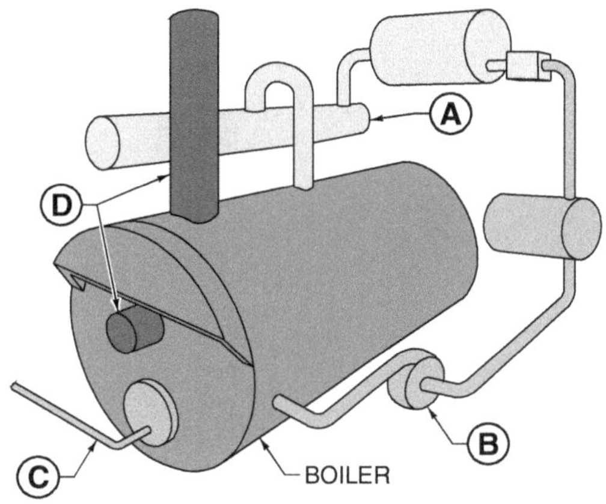

pH Scale

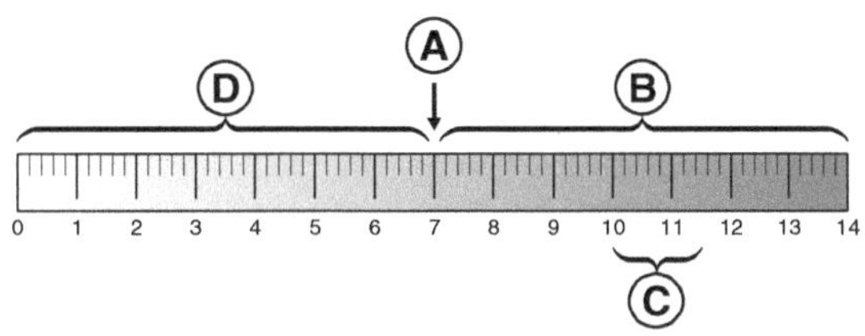

______________ 1. Acidic

______________ 2. Alkaline

______________ 3. Neutral

______________ 4. pH commonly required for boiler water

Heating Systems—Steam

_______________ **1.** Boiler

_______________ **2.** Steam flow

_______________ **3.** Steam trap

_______________ **4.** Heating unit

_______________ **5.** Main steam stop valve

_______________ **6.** Feedwater pump

_______________ **7.** Steam bubbles

_______________ **8.** Condensate flow

_______________ **9.** Condensate return line

_______________ **10.** Feedwater flow

_______________ **11.** Main steam line

_______________ **12.** Branch line

_______________ **13.** Steam header

_______________ **14.** Condensate receiver tank

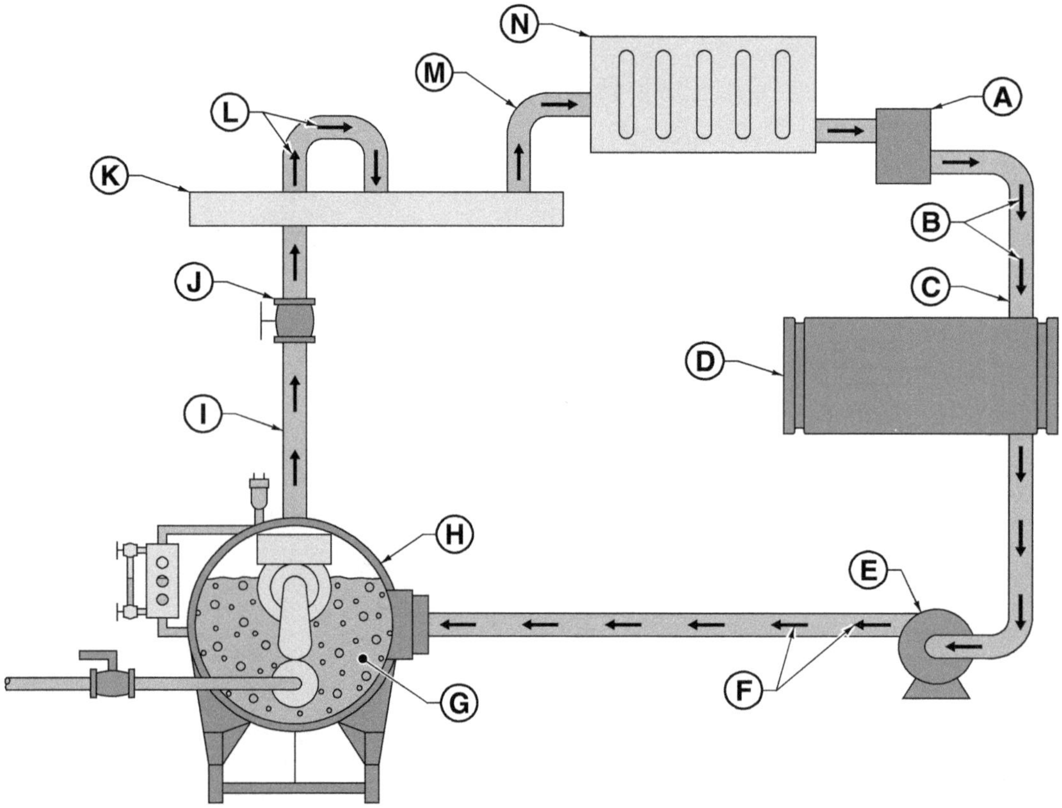

Heating Systems—Hot Water

______________ **1.** Hot water flow

______________ **2.** Heating unit

______________ **3.** Compression tank

______________ **4.** Boiler full of water

______________ **5.** Branch line

______________ **6.** Backflow preventer

______________ **7.** Circulating pump

______________ **8.** Makeup water supply line

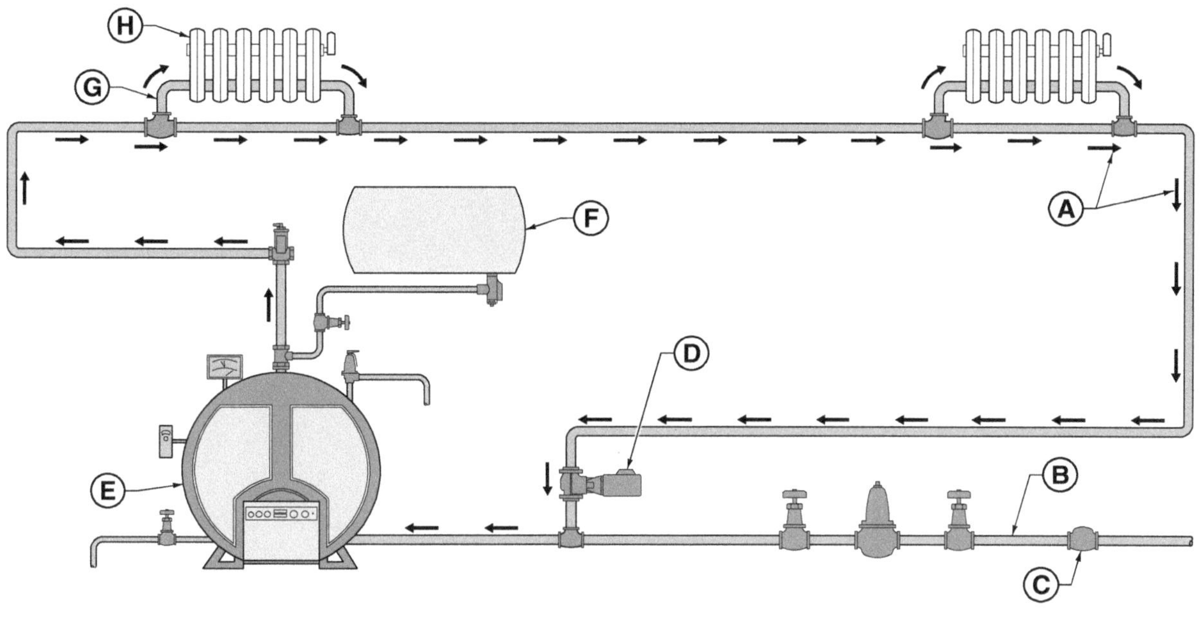

Water Columns

_____ 1. Water

_____ 2. Cross tees

_____ 3. Try cocks

_____ 4. Gauge glass blowdown valve

_____ 5. Vent

_____ 6. Top of boiler

_____ 7. Water column

_____ 8. Water level

_____ 9. Top isolation valve

_____ 10. Gauge glass

_____ 11. NOWL

_____ 12. Alarm sensor switch enclosure

_____ 13. Bottom isolation valve

_____ 14. Water column blowdown valve

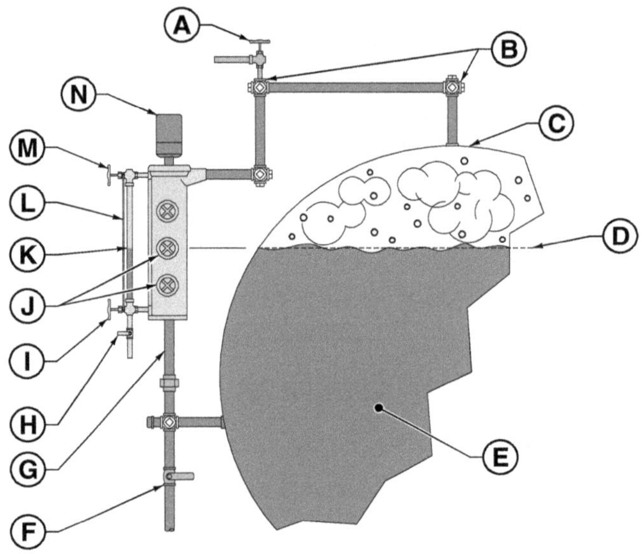

Feedwater Valves

____________________	**1.** Boiler	____________________	**4.** Stop valves
____________________	**2.** Feedwater pump	____________________	**5.** Main feedwater line
____________________	**3.** Check valves		

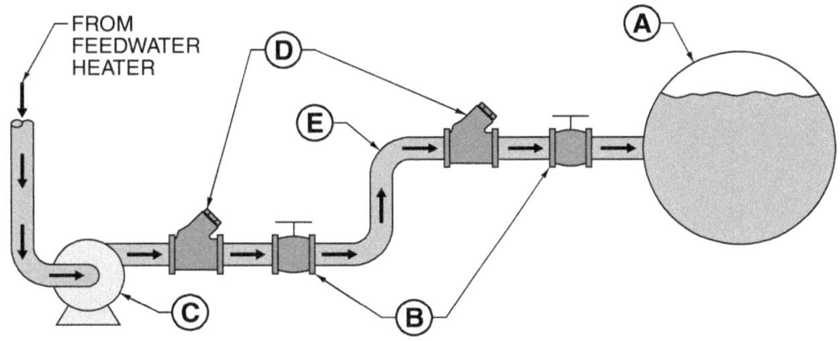

Natural Gas Accessories

____________________	**1.** Pilot gas regulator	____________________	**6.** Manual shutoff
____________________	**2.** Coupling	____________________	**7.** Motorized valves
____________________	**3.** Regulator	____________________	**8.** High gas pressure switch
____________________	**4.** Low gas pressure switch	____________________	**9.** Pilot solenoid valve
		____________________	**10.** Vent valve
____________________	**5.** Manual pilot shutoff valve	____________________	**11.** Butterfly valve

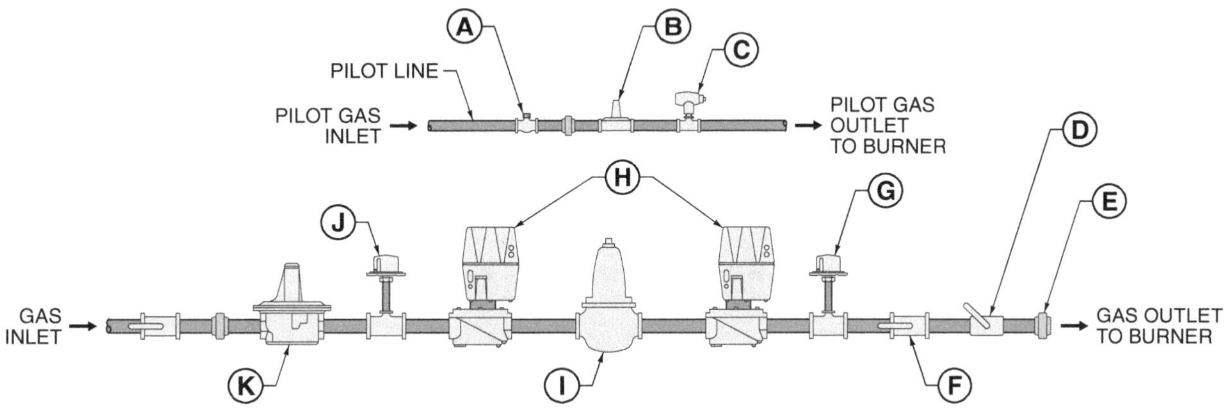

Mechanical Draft—Forced

____________	**1.** Forced draft fan	____________	**6.** Airflow
____________	**2.** Boiler drum	____________	**7.** Furnace
____________	**3.** Air entering furnace	____________	**8.** Outlet damper
____________	**4.** Gases of combustion	____________	**9.** Inlet damper
____________	**5.** Breeching	____________	**10.** Chimney

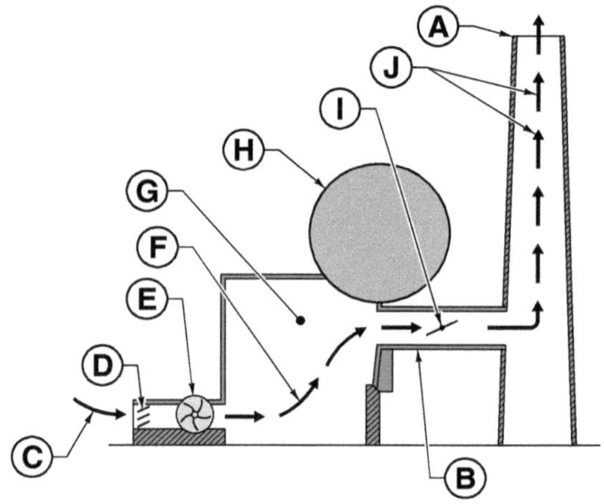

Activity: Boiler Classification

Answer the following questions about the classification of boilers.

1. Why might the AHJ lower the MAWP of a boiler?

2. Would it be legal to make a boiler that was designed as a low-pressure boiler into a high-pressure boiler, and what might happen if this were done?

3. What are some differences between watertube and firetube boilers?

Activity: Heating Systems

A steam heating system is used in a facility.

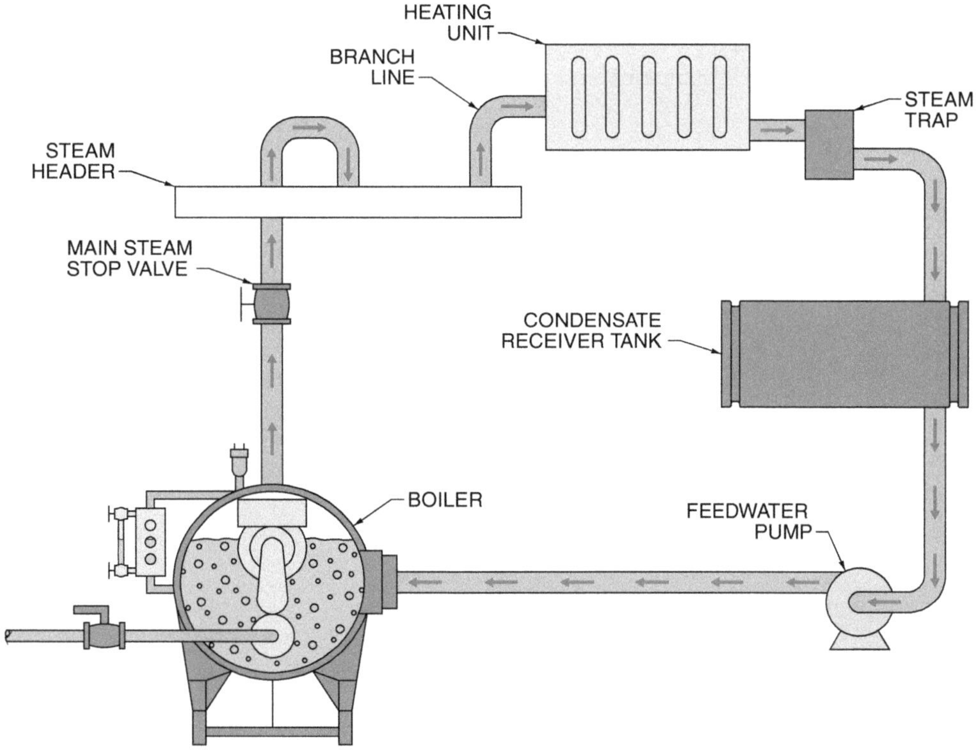

1. What changes occur to the water/steam as it moves through the system?

2. Indicate on the drawing where the latent heat is added to the water/steam by the system and where is it given up by the water/steam.

A hot water heating system is used in a facility.

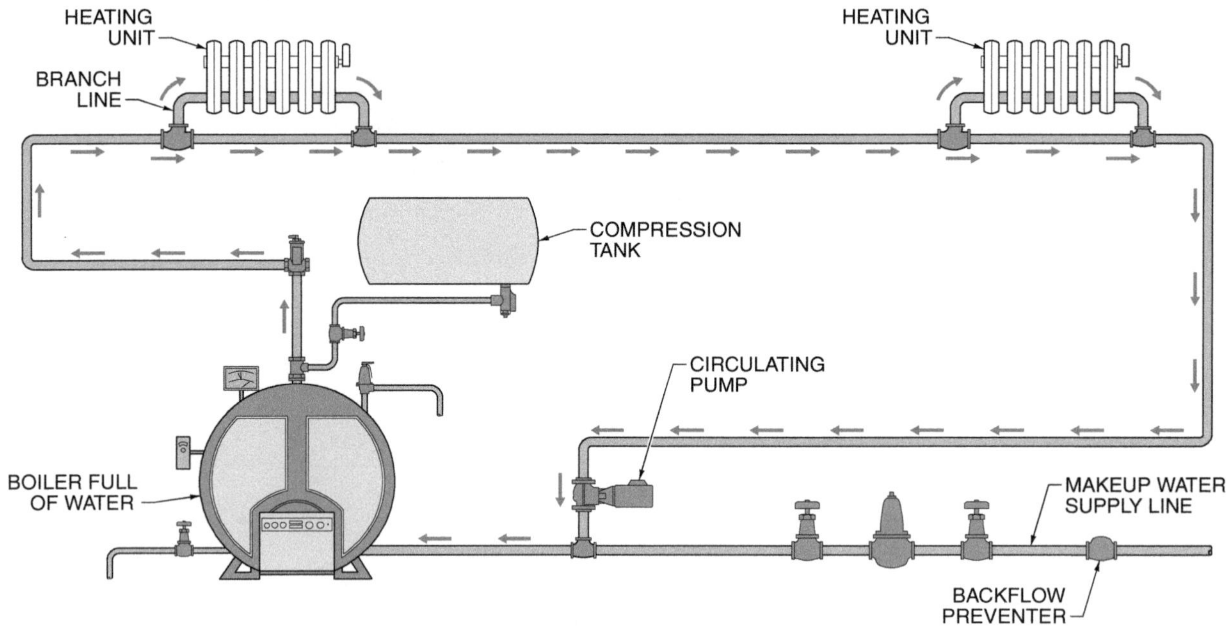

3. Does the water change states as it moves through the system?

4. Indicate on the drawing where the sensible heat is added to the water by the system and where it is given up by the water.

Activity: Low Water Fuel Cutoff

Water from a low water fuel cutoff is blown down and the burner is still firing.

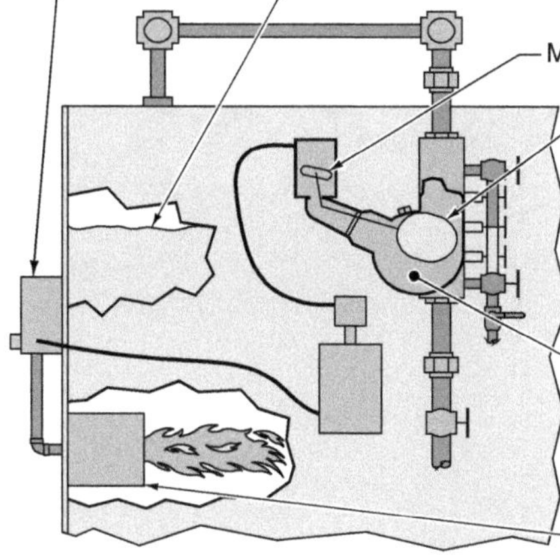

1. What is the problem in this situation?

2. How could this problem be repaired?

3. How could this problem be prevented from reoccurring?

4. What serious situation could result if this problem is not repaired?

Activity: Steam Traps

Temperature readings are taken at two points in the pipelines near a steam trap.

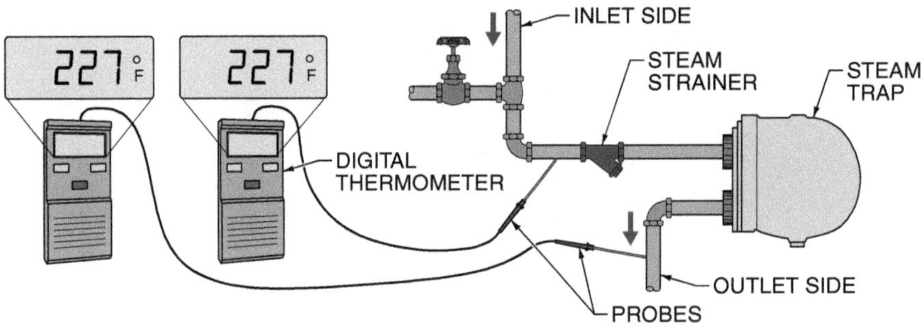

1. What is the problem in this situation?

2. What are two possible causes of this problem?

3. How could this problem be repaired?

4. How could this problem be prevented from reoccurring?

Activity: Natural Gas Line

A natural gas line feeds the fuel system of a boiler.

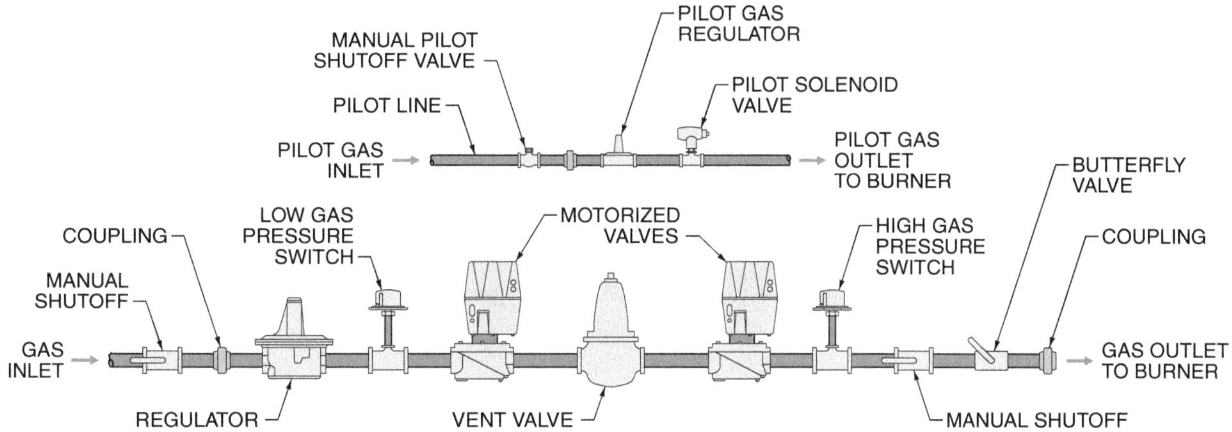

1. Why is gas easier to burn than fuel oil?

2. How is a low gas pressure cutoff tested?

3. What might happen to the boiler flame if the gas pressure is too low?

4. Why are there two motorized valves on the gas line?

5. What would happen if the solenoid valve on the pilot gas line were defective and would not open?

Activity: Flame Scanner Failure

The flame scanner from a boiler furnace is removed and tested.

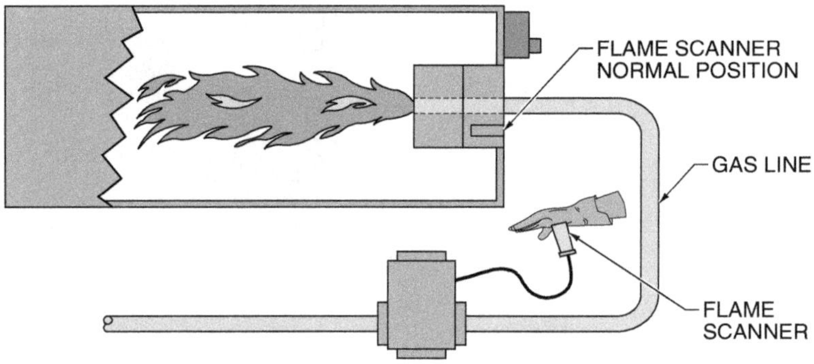

1. Describe how a flame scanner is tested.

2. What is the problem in this situation?

3. How could this problem be repaired?

4. What serious situation could result if this problem were not repaired?

Activity: Water Column and Gauge Glass Blowdown

The gauge glass and water column of a boiler system are blown down. The water flows sluggishly into and out of the gauge glass and water column.

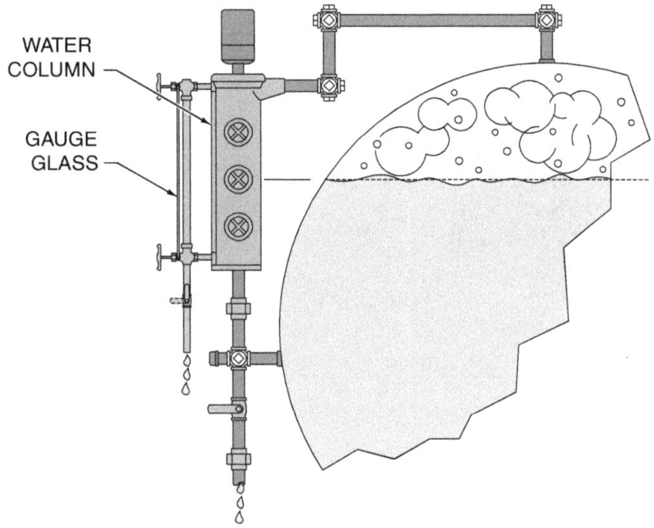

1. What is the most likely cause of this problem?

2. How could this problem be repaired?

3. How could this problem be prevented from reoccurring?

4. How often should the water column and gauge glass be blown down?

Activity: Boiler Blowdown

A boiler is operating under normal conditions.

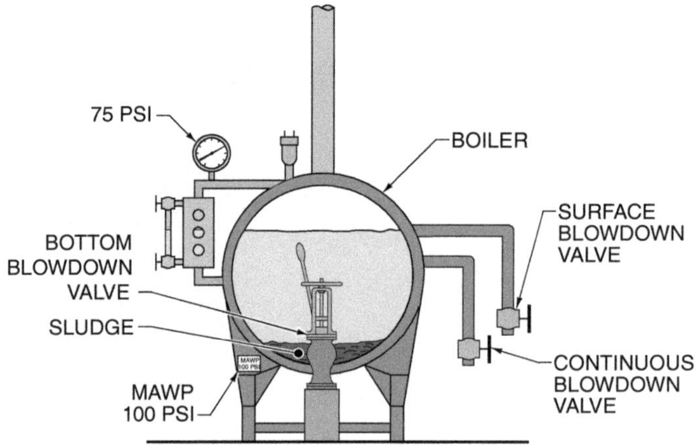

1. When should the boiler be blown down?

2. What determines how much water should be removed from the boiler?

3. Except when a boiler is being completely drained, what three conditions should be met before the boiler is blown down?

4. What could happen if these conditions are not met when this boiler is blown down?

5. When a boiler is emptied using the blowdown valves and has been washed out, what safety precaution should be taken with the blowdown valves?

6. Why are there two blowdown valves?

Activity: Low Water in the Boiler

A boiler has no water in the gauge glass. The steam pressure gauge reading is within the normal range, and the boiler is still firing.

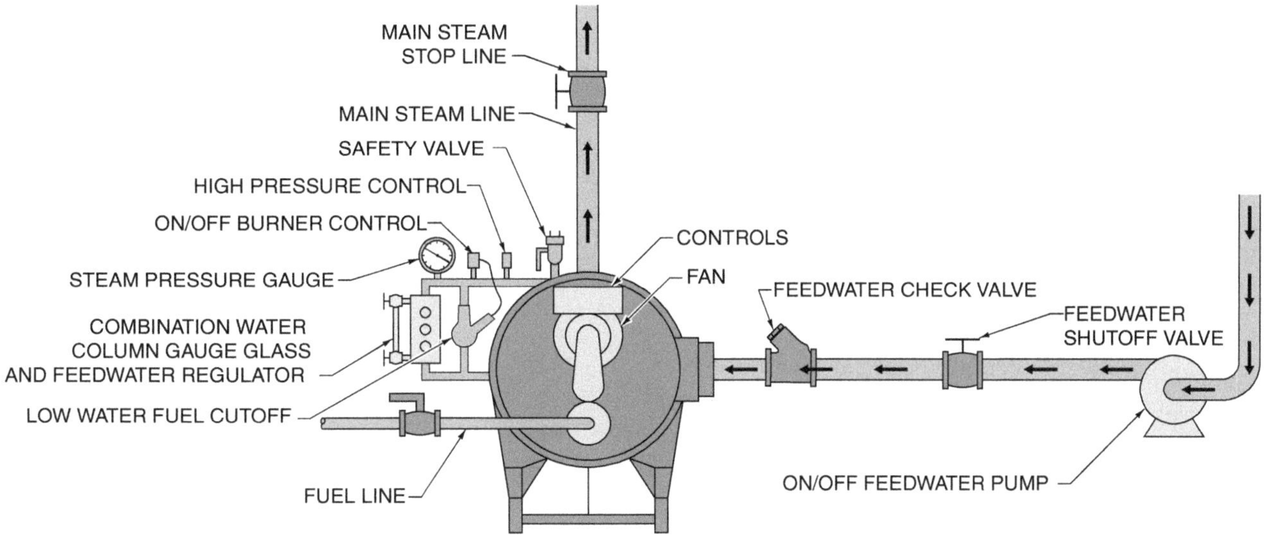

1. What should be done first?

2. What should be done next?

3. Which electrical devices might have failed and caused the problem?

4. What are possible causes of these failures?

5. Which system should be completely inspected to investigate these possible causes?

6. How could this problem be prevented from reoccurring?

Activity: High Water in the Boiler

A boiler has a gauge glass full of water. The steam pressure gauge reading is within the normal range, and the boiler is still firing.

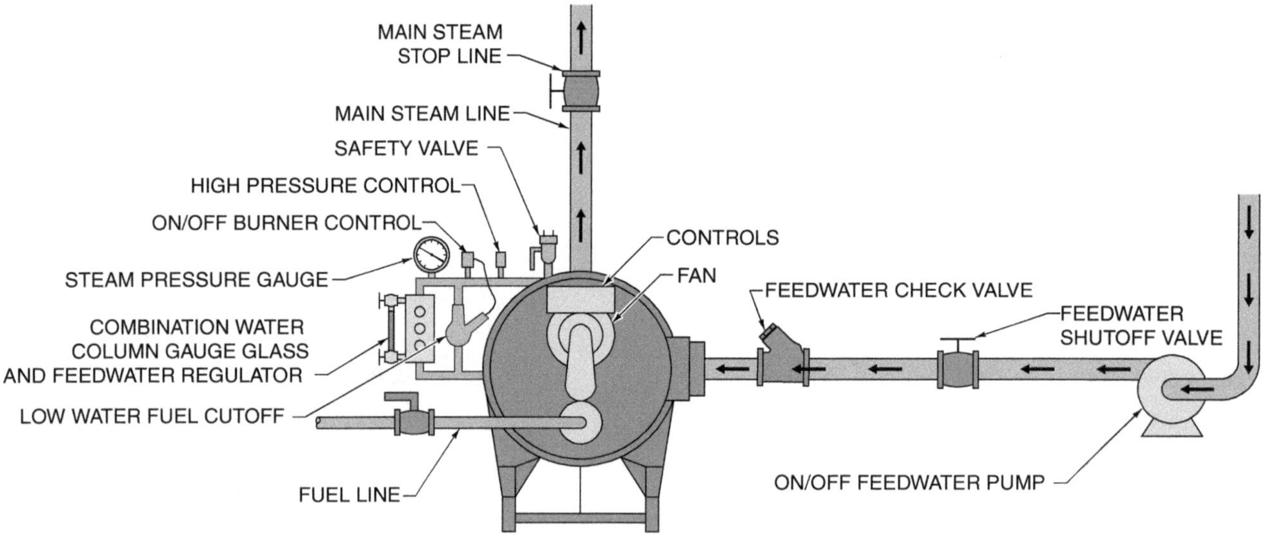

1. What should be done first?

2. What should be done next?

3. Which device might have failed and caused the problem?

4. What are possible causes of this failure?

5. Which system should be completely inspected to investigate these possible causes?

6. How could this problem be prevented from reoccurring?

Activity: Boiler Overpressure

A boiler steam pressure gauge reads 15 psi over the MAWP, but the safety valve is not open, and the boiler is still firing.

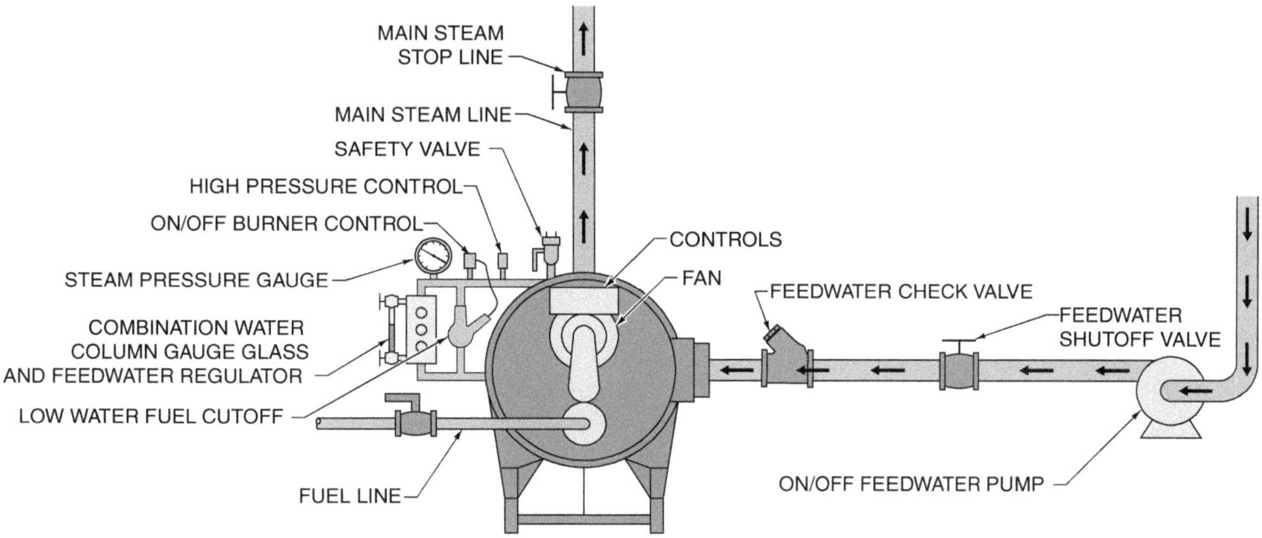

1. What should be done first? What should not be done?

2. What should be done next?

3. Which devices failed and caused the problem?

4. What are possible causes of these failures?

5. How could this problem be corrected?

6. How could this problem be prevented from reoccurring?

Activity: Boiler Shutdown and Startup

Answer the following questions about the shutdown and startup of a boiler.

1. What might happen if the vent were not opened during a boiler shutdown?

2. List all boiler components that must be locked out when the boiler is shut down and opened for its annual inspection.

3. What damage might occur if a cold boiler were started and kept on high fire?

4. Why must steam piping be warmed slowly during boiler startup?

COMFORT

SECTION 10.1

REVIEW QUESTIONS

Name ______________________________ Date ______________

True-False

T	F	**1.** When clothed, the human body is comfortable at an air temperature of approximately 70°F to 75°F.
T	F	**2.** Moisture is not always present in the air.
T	F	**3.** Higher humidity slows the evaporation rate, and lower humidity increases the evaporation rate.
T	F	**4.** Air pressure does not vary inside a building.
T	F	**5.** Air circulation can cause temperature stratification.
T	F	**6.** Indoor air can become much more polluted than outside air.

Multiple Choice

______________ **1.** ___ is the amount of moisture present in the air.

- A. Saturation
- B. Stratification
- C. Temperature
- D. Humidity

______________ **2.** ___ is the air temperature below which moisture condenses.

- A. Dewpoint
- B. Melting point
- C. Absolute humidity
- D. Wet-bulb temperature

______________ **3.** At normal room temperatures, comfortable relative humidity levels are between about ___.

- A. 5% and 25%
- B. 30% and 50%
- C. 60% and 70%
- D. 85% and 95%

______________ **4.** A ___ is a cover for an opening of ductwork from a building space that allows air to return to the HVAC system.

- A. register
- B. diffuser
- C. grille
- D. baffle

______________ **5.** ___ is the process of introducing fresh outdoor air into a building.
A. Ventilation
B. Air conditioning
C. Circulation
D. Filtration

______________ **6.** ___ is the natural rising of hot air up the center of a building through elevator shafts, stairwells, and service columns.
A. Temperature stratification
B. Ventilation
C. Air circulation
D. Stack effect

Completion

______________ **1.** A(n) ___ is a cover for an opening of ductwork into a building space that includes movable vanes to control the amount and direction of the supply airflow.

______________ **2.** ___ is the process of removing particulate matter from air.

______________ **3.** ___ occurs when a person is not able to sense a difference between themselves and the surrounding air.

______________ **4.** Indoor air ___ is a description of the type and quantity of contaminants in indoor air.

______________ **5.** ___ humidity is the ratio of the amount of water vapor in air to the maximum moisture capacity of the air at a certain temperature.

______________ **6.** ___ air is air with high concentrations of carbon dioxide and/or other vapor pollutants.

SYSTEM OPERATION

SECTION **10.2**

REVIEW QUESTIONS

Name ______________________________ **Date** ____________________

True-False

T F **1.** A closed-loop control system continuously checks the outcome of its actions to determine when operations should change.

T F **2.** Variables include sensors, controllers, actuators, and human-machine interfaces (HMIs).

T F **3.** With the widespread use of building automation systems, control system accuracies of ±0.5°F are possible.

Multiple Choice

__________________ **1.** A(n) ___ is a device that makes decisions to change some aspect of a system based on sensor information and internal programming.

- A. human-machine interface
- B. actuator
- C. variable
- D. controller

__________________ **2.** ___ are used to change system characteristics, such as dampers for regulating airflow and gas valves and electric heating elements for regulating heating.

- A. Actuators
- B. Controllers
- C. Variables
- D. Setpoints

__________________ **3.** A normally open (NO) device is a device that is open when the signal causing it to move is ___.

- A. present
- B. input into an HMI
- C. at its lowest level
- D. at its highest level

Completion

__________________ **1.** A(n) ___ is an interface terminal that allows an individual to access and respond to system information.

__________________ **2.** A(n) ___ is a set of adjustable metal blades used to control the amount of airflow between two spaces.

__________________ **3.** A(n) ___ is the desired value to be maintained by a system.

FORCED-AIR SYSTEM CONFIGURATIONS

SECTION **10.3**

REVIEW QUESTIONS

Name ______________________ **Date** ______________

True-False

T F **1.** A building space may require cooling even when the outdoor weather is cool.

T F **2.** Outside-air and return-air dampers maintain equal but opposite positions.

T F **3.** Constant-air-volume (CAV) air handlers are the most common air handlers installed in new commercial and industrial buildings.

T F **4.** Two methods of adding conditioned air to indoor building spaces include the constant-volume method and the variable-volume method.

T F **5.** Most HVAC systems are open-loop systems.

Multiple Choice

______________ **1.** ___ air is the air from within a building space that is drawn back into a forced-air HVAC system to be exhausted or reconditioned.

A. Return
B. Exhaust
C. Outside
D. Mixed

______________ **2.** The disadvantage of CAV air handlers is that the fans always operate at ___ power.

A. minimum
B. 50%
C. 75%
D. maximum

______________ **3.** ___ is a cooling strategy that adds cool outside air to supply air.

A. Ventilation
B. Economizing
C. Filtration
D. Circulation

______________ **4.** If the outside-air temperature is so cold that the frigid air would freeze heating pipes or damage other components, the HVAC system temporarily shuts down, which is called ___.

A. automatic shutdown
B. temperature control
C. freezestat control
D. economizing

________________ **5.** A ___ is an area within a building that shares HVAC requirements.
A. room
B. zone
C. floor
D. house

Completion

________________ **1.** A(n) ___ HVAC system is a system that distributes conditioned air throughout a building in order to maintain the desired conditions.

________________ **2.** An air ___ is a device that conditions and distributes air to spaces in a building.

________________ **3.** ___ air is newly conditioned mixed air that is distributed to a building space.

________________ **4.** A(n) ___ unit is a device that is located close to a zone and heats or cools air flowing through it.

________________ **5.** A(n) ___ box is a terminal unit located at a building zone that provides heating and airflow as needed to maintain the desired setpoints within the building zone.

HVAC SYSTEM COMPONENTS

SECTION 10.4

REVIEW QUESTIONS

Name ______________________________ **Date** ____________________

True-False

T F **1.** The two most common variables for determining fan size are fan capacity and static pressure.

T F **2.** Static pressure is the pressure exerted by airflow in a direction parallel to flow.

T F **3.** Filter efficiency is related to the sizes of the particulates that are caught in the filter compared to those that pass through.

T F **4.** The most common type of airflow station is a pitot tube station.

T F **5.** The C port of a three-way valve allows water to exit the valve.

Multiple Choice

______________ **1.** A ___ is a device that measures and transmits temperature.
- A. thermostat
- B. temperature sensor
- C. hygrometer
- D. thermocouple

______________ **2.** A ___ is a switch that activates at humidity levels either above or below a certain setpoint.
- A. hygrometer
- B. hygroscopic element
- C. humidimeter
- D. humidistat

______________ **3.** The most common hygrometers measure ___.
- A. percent relative humidity (%RH)
- B. pressure
- C. wet-bulb temperature
- D. airflow

______________ **4.** A(n) ___ is a sensor that measures the velocity of the air in a duct system.
- A. differential pressure switch
- B. flow switch
- C. airflow station
- D. pressure sensor

_______________ **5.** A(n) ___ is a set of adjustable metal blades used to control the amount of airflow between two spaces.
A. fan
B. damper
C. grille
D. airflow station

Completion

_______________ **1.** A(n) ___ is a device that removes moisture from air by causing the moisture to condense.

_______________ **2.** Fan ___ is the volume of air a fan can move in a given time.

_______________ **3.** A heating ___ is a finned heat exchanger that adds heat to air.

_______________ **4.** An air ___ is a porous device that removes particulates from air.

_______________ **5.** A cooling ___ is a finned heat exchanger that removes heat from air.

CONTROL SIGNALS

SECTION **10.5**

REVIEW QUESTIONS

Name ______________________________ **Date** ____________________

True-False

T F **1.** Control signals can only be conveyed as electronic signals or radio frequencies.

T F **2.** A digital signal is typically conveyed with a change in voltage.

T F **3.** Pneumatic control systems commonly use a range of 10 psig to 20 psig.

Multiple Choice

__________ **1.** A digital signal is a signal that has only ___ possible state(s).
- A. one
- B. two
- C. three
- D. four

__________ **2.** Any pair of two different voltages can be used to send digital signals, though they are commonly DC voltages of ___ VDC or less.
- A. 10
- B. 24
- C. 50
- D. 63

__________ **3.** A ___ is a standardized set of rules and procedures for the exchange of information between two connected devices.
- A. guidebook
- B. manual
- C. protocol
- D. network

Completion

__________ **1.** A(n) ___ is a changing characteristic used to communicate information between control devices.

__________ **2.** A(n) ___ signal is a signal that has a continuous range of possible values between two points.

__________ **3.** The most common electrical properties used to convey analog signals are voltage, current, and ___.

CONTROL SYSTEMS

SECTION **10.6**

REVIEW QUESTIONS

Name ______________________________ **Date** ______________

True-False

T F **1.** A pressure-relief valve is located before a pressure regulator to ensure that high air pressure does not damage the equipment downstream.

T F **2.** Most controller designs are based on a balancing of pressures with a fulcrum (balancing point) determined by setpoint adjustment screws.

T F **3.** Direct digital controls (DDCs) send and receive variable voltage and current signals that operate pneumatic or electric actuators.

T F **4.** Though security systems are the most common type of system to automate, other implementations include HVAC, lighting, electrical system, access, and elevator controls.

Multiple Choice

______________ **1.** A pneumatic control air supply consists of two compressors and a(n) ___ that controls the compressors as system pressure changes.

A. thermostat
B. on/off pressure switch
C. pressure-relief valve
D. pneumatic-electric switch

______________ **2.** A(n) ___ is a switch powered by a solenoid that moves a plunger, opening or closing an air line.

A. electric-pneumatic switch
B. pneumatic controller
C. pneumatic actuator
D. pneumatic-electric switch

______________ **3.** A pneumatic ___ is a device that uses air pressure to position HVAC components.

A. thermostat
B. operator
C. switch
D. actuator

Completion

______________ **1.** A(n) ___ is an analog device that converts some climate variable into a proportional air pressure signal, typically between 3 psig and 15 psig.

__________________ **2.** A(n) ___ system is a control system that uses low voltage and current to send variable signals from sensors and controllers to controlled devices.

__________________ **3.** A(n) ___ system is a control system that uses digitally encoded messages shared on a common communication network to distribute information between control devices.

__________________ **4.** A(n) ___ control system is a control system that uses compressed air to send variable signals from sensors and controllers to controlled devices.

HVAC SYSTEM MAINTENANCE

SECTION **10.7**

REVIEW QUESTIONS

Name ______________________________ **Date** ______________

True-False

T F **1.** Cleaning, maintenance, remodeling, and building repairs affect indoor air quality.

T F **2.** Leaking valves display a temperature difference from one side to the other when closed.

T F **3.** Other than a yearly operational check, DDC systems should require very little maintenance.

Multiple Choice

______________ **1.** A clogged filter is indicated by ___.
   A. an accumulation of dirt in the building
   B. mold and bacteria growth
   C. dirt on the downstream side of the filter
   D. an increased pressure drop across the filter

______________ **2.** A common problem with dampers is ___.
   A. broken louvers
   B. unexpected movement
   C. rapid positioning
   D. loose fasteners that connect the dampers to connecting rods

______________ **3.** The primary focus of pneumatic system maintenance is the ___.
   A. compressor
   B. filter
   C. supply-air damper
   D. outside-air damper

Completion

______________ **1.** ___ relative humidity can cause respiratory problems and irritate mucous membranes.

______________ **2.** Construction areas should be sealed and kept under ___ pressure, and return air should be exhausted outside.

______________ **3.** A filter that is clogged and needs replacement will likely be blackened on the ___ (low-pressure) side from particulates trapped in the filter material.

HVAC SYSTEM TROUBLESHOOTING

SECTION **10.8**

REVIEW QUESTIONS

Name ______________________________ Date ______________

True-False

T F **1.** The too hot/cold call is the most common HVAC system complaint encountered.

T F **2.** Low carbon dioxide levels indicate inadequate ventilation.

T F **3.** Direct digital control (DDC) system problems are commonly located and repaired from a computer terminal.

Multiple Choice

______________ **1.** In a pneumatic control system, the ___ becomes brittle and may crack or pop off fittings as it ages.
- A. plastic tubing
- B. signal wiring
- C. actuator
- D. diaphragm

______________ **2.** ___ cycling is a frequent on/off operation of equipment due to improper feedback.
- A. Temporary
- B. Intermittent
- C. Sporadic
- D. Short

______________ **3.** The primary disadvantage of building automation systems is that ___ than with other control systems.
- A. the systems are much harder to control
- B. the design and programming are more complex
- C. short cycling occurs much more frequently
- D. more maintenance is required

Completion

______________ **1.** To troubleshoot IAQ problems, first the ___ and humidity in the space are checked.

______________ **2.** To prevent ___, diffusers are adjusted and airflow obstructions are removed to ensure that air circulates through the occupied zone.

______________ **3.** To check a sensor, its ___ signal is measured.

HVAC SYSTEMS

Chapter 10

ACTIVITIES

Name ______________________ **Date** ______________

Air Handlers

________ **1.** Supply-air damper	________ **10.** Differential pressure sensor
________ **2.** Temperature sensor	________ **11.** Fan
________ **3.** Outside air	________ **12.** Outside-air damper
________ **4.** Exhaust-air damper	________ **13.** Supply air
________ **5.** Filter	________ **14.** Mixed air
________ **6.** Humidifier	________ **15.** Return-air damper
________ **7.** Airflow station	________ **16.** Heating coil
________ **8.** Cooling coil	________ **17.** Exhaust air
________ **9.** Return air	

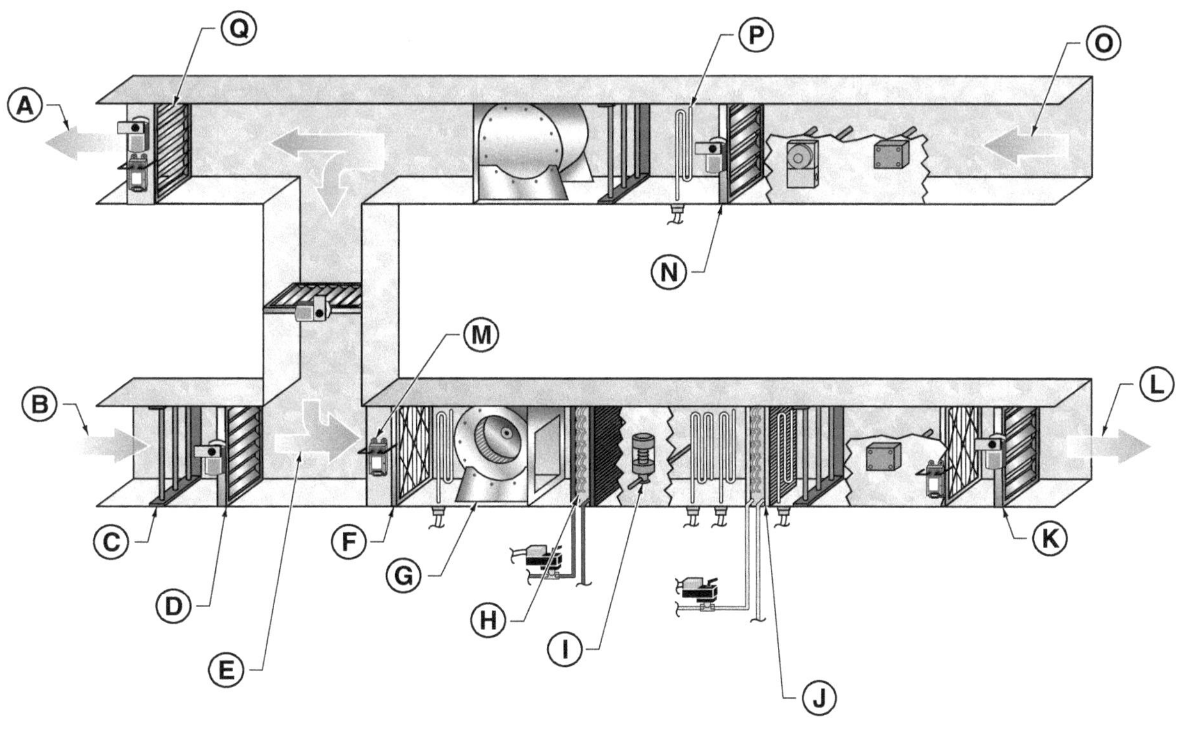

Variable-Air-Volume (VAV) Box

_______________ **1.** Heating coil

_______________ **2.** Air to building space

_______________ **3.** Differential pressure sensor

_______________ **4.** Airflow station

_______________ **5.** Air from air handler

_______________ **6.** Damper

_______________ **7.** Temperature sensor

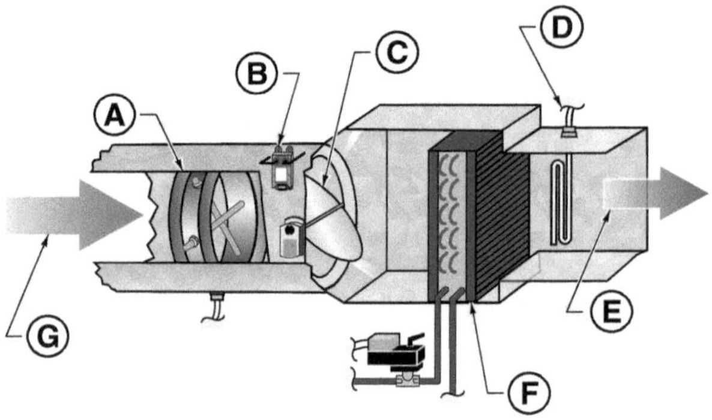

Analog Signals

_______________ **1.** What is the range of possible current signals?

_______________ **2.** What is the maximum temperature that can be indicated by this signal?

_______________ **3.** What is the minimum temperature that can be indicated by this signal?

_______________ **4.** What is the approximate temperature indicated by a 10 mA signal?

_______________ **5.** What is the approximate temperature indicated by a 17 mA signal?

_______________ **6.** What is the approximate current signal for 200°F?

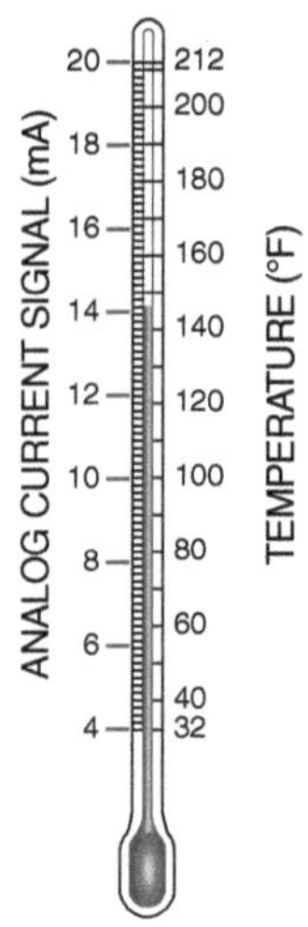

Fans

______________________ **1.** Radial flow, direct drive

______________________ **2.** Radial flow, indirect drive

______________________ **3.** Axial flow, direct drive

______________________ **4.** Axial flow, indirect drive

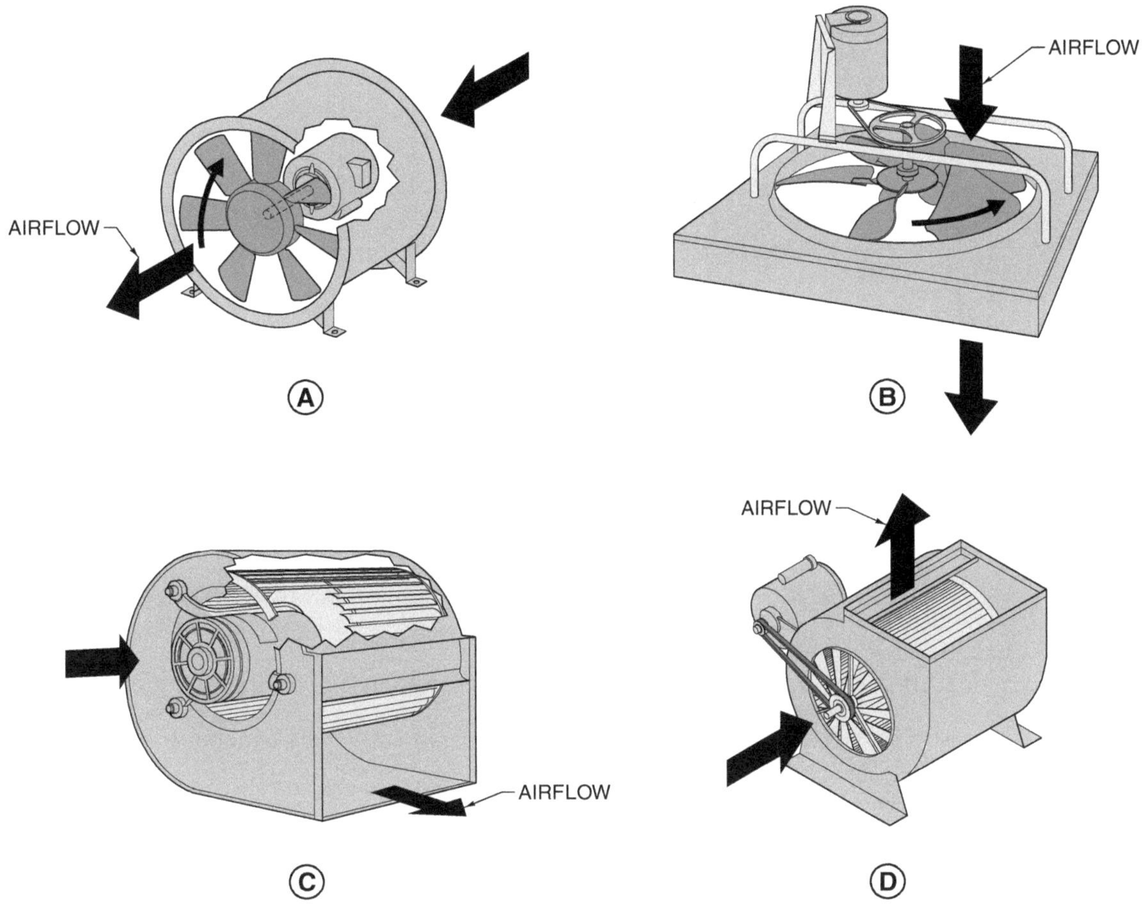

Fan Capacity

Use the fan capacity graph to answer the following questions.

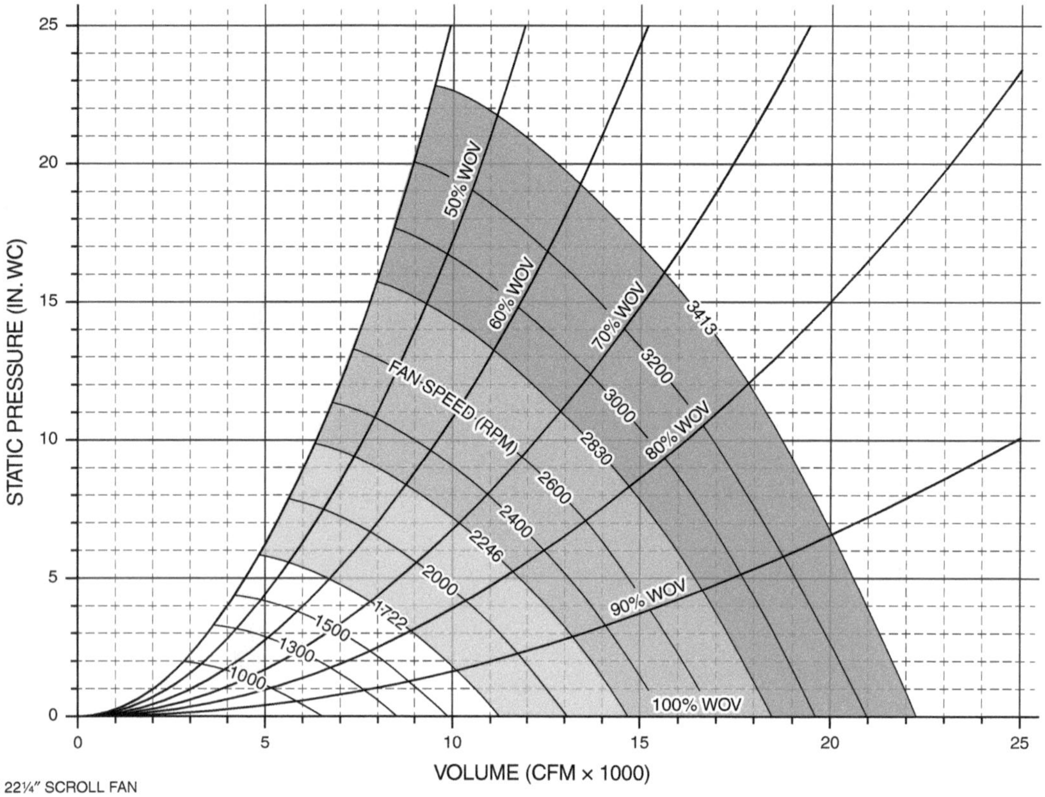

* FOR 22¼″ SCROLL FAN

________________ **1.** At 100% wide-open volume (WOV), this fan operating at 1300 rpm moves about ___ cfm of air.

________________ **2.** If the fan rpm is increased to 2600 rpm, the fan moves about ___ cfm of air.

________________ **3.** The fan is connected to ductwork that decreases the WOV to about 80%. At 2600 rpm, the static pressure is about ___ in. WC.

________________ **4.** The fan is operating at 3000 rpm and 70% WOV when a damper closes, increasing the resistance to about 60% WOV. This increases the static pressure by about ___ in. WC.

________________ **5.** If the fan must move a volume of 10,000 cfm and the system has a resistance of about 70% WOV, the fan should operate at about ___ rpm.

________________ **6.** If the fan must move a volume of 14,000 cfm and the system has a resistance of about 90% WOV, the fan should operate at about ___ rpm.

Pneumatic Control Systems

______________ **1.** Pneumatic-electric switch

______________ **2.** Damper actuator

______________ **3.** Compressor

______________ **4.** Low-pressure gauge

______________ **5.** Pressure regulator

______________ **6.** Receiver

______________ **7.** Thermostat

______________ **8.** Variable pressure

______________ **9.** Pneumatic valve

______________ **10.** High-pressure gauge

______________ **11.** Refrigerated air dryer

______________ **12.** On/off pressure switch

______________ **13.** Constant pressure

______________ **14.** Pressure-relief valve

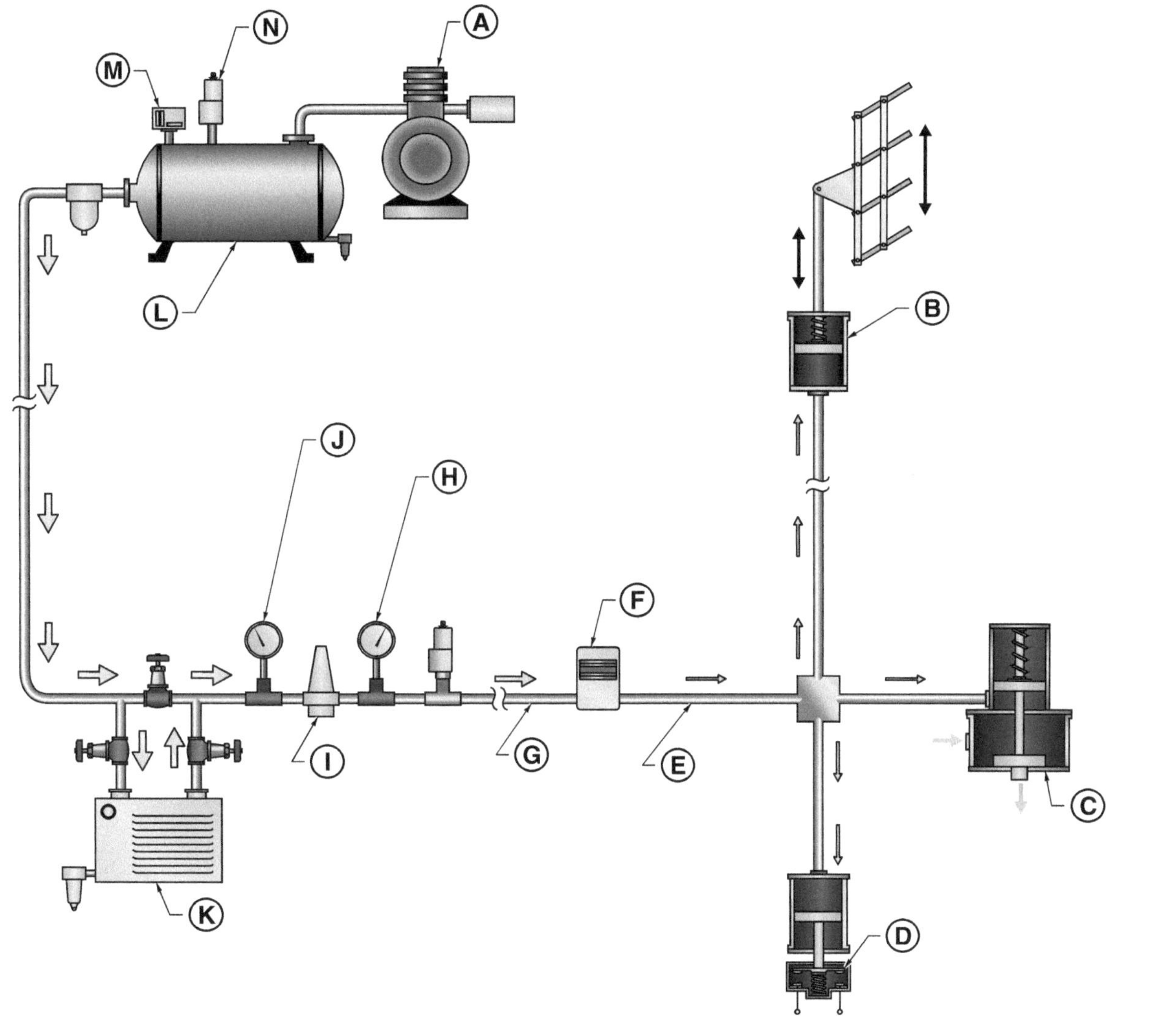

Pneumatic Actuators

_______________ **1.** Piston

_______________ **2.** Diaphragm

_______________ **3.** Bellows

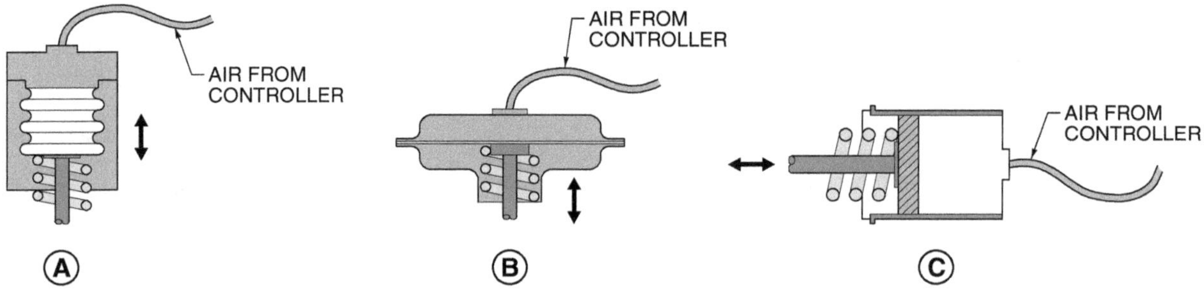

Activity: Air Handler Operation

An air handler must be adjusted to cool a building space when it is hot outside.

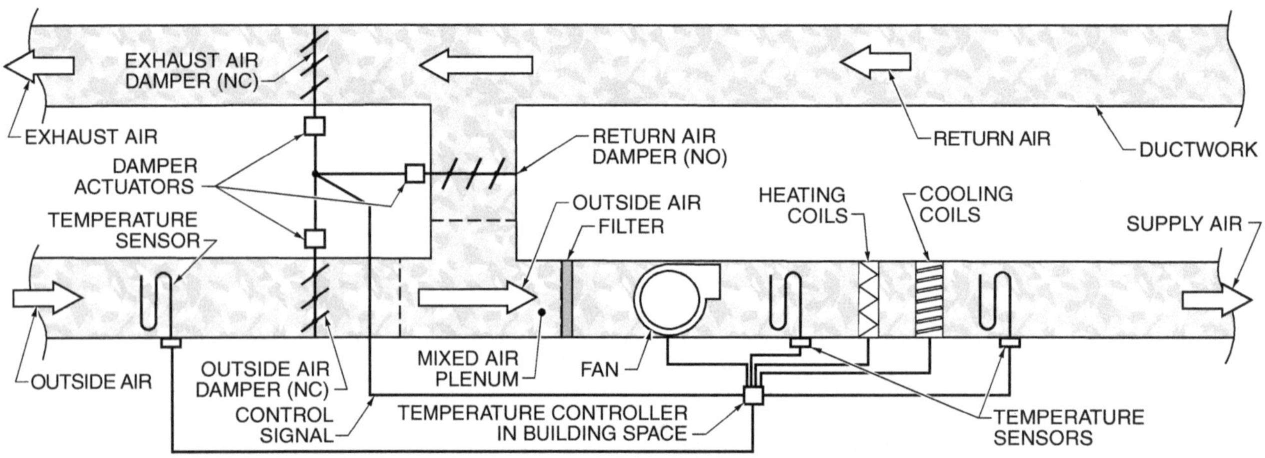

1. How should the outside, exhaust, and return air dampers be positioned?

2. Which of the coils will be operating?

The air handler must be adjusted to cool the building space when it is cool outside.

3. How should the outside, exhaust, and return air dampers be positioned?

4. Which of the coils will be operating?

The air handler must be adjusted to heat the building space when it is cold outside.

5. How should the outside, exhaust, and return air dampers be positioned?

6. Which of the coils will be operating?

7. What problem would occur if the outside air damper were locked in the closed position?

8. What problem would occur if the return air damper were stuck closed?

9. What maintenance would this unit require?

Activity: Variable-Air-Volume Box

A variable-air-volume (VAV) box is used in a system to control the temperature of a zone.

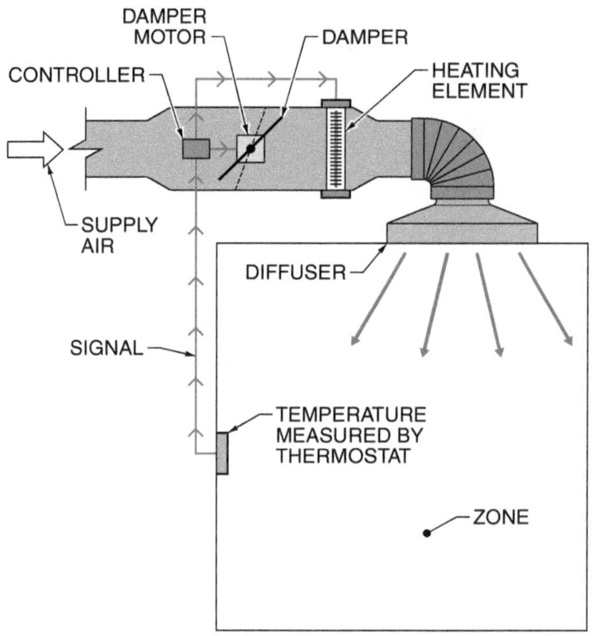

1. Why is this VAV box considered a closed-loop system?

2. What might occur if the thermostat were damaged and would not send a signal to the controller?

3. Which position should the damper be in when heating is required?

4. Which position should the damper be in when cooling is required?

5. What will happen to a sensor signal that is sent along a signal wire that is too long or has a loose connection point? (Think about Ohm's law.)

The zone includes a bookcase that is placed in front of the thermostat.

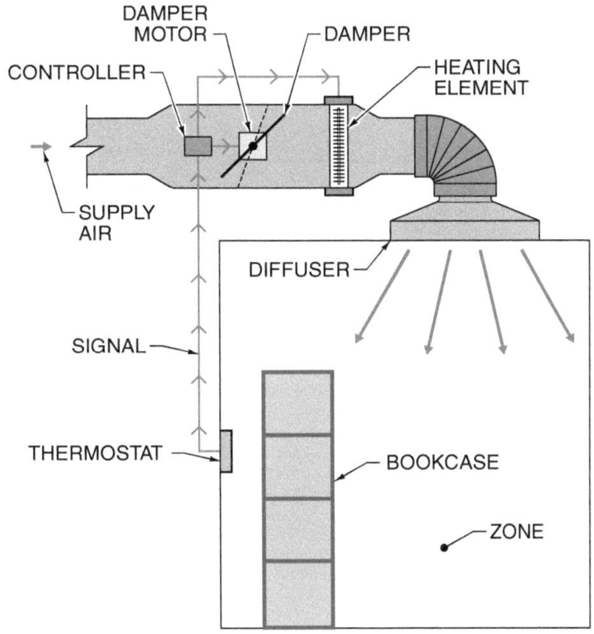

6. What is the problem in this situation?

The zone includes an office partition that divides the room.

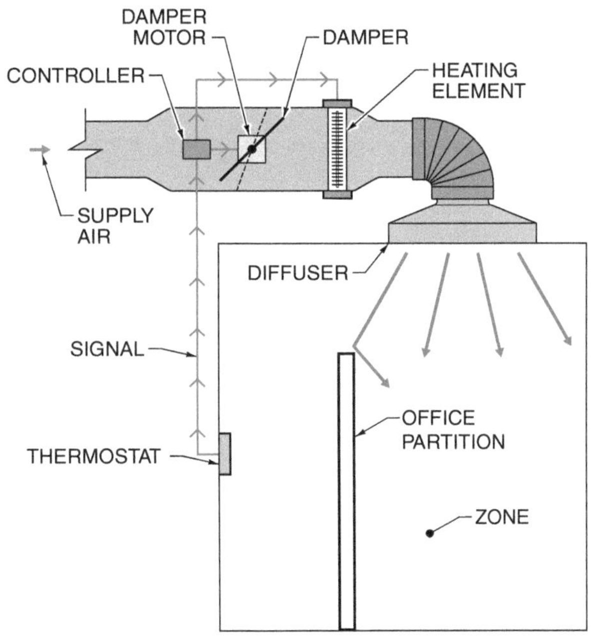

7. What is the problem in this situation?

Activity: Building Pressures

Chemicals are stored in a closet with a fan that exhausts chemical fumes outside.

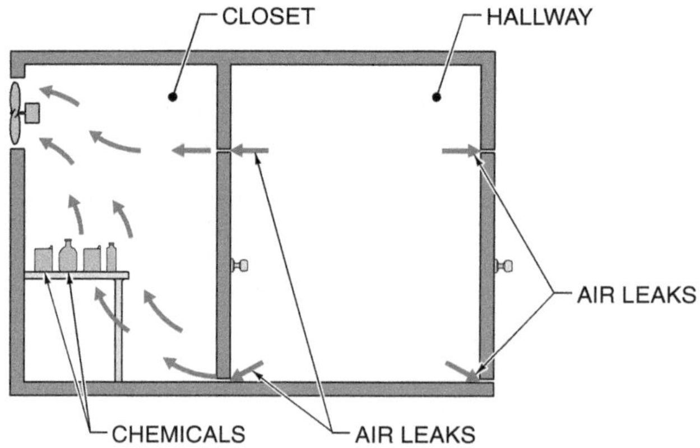

1. What is the air pressure in the closet relative to the air pressure in the hallway?

2. What would happen to the air pressure in the closet if the exhaust fan failed?

3. What problem might develop in the hallway if the exhaust fan failed?

4. What basic maintenance would a belt-driven exhaust fan require?

5. Should the exhaust air from this chemical closet be added to the overall return air for an air handler?

6. What problem might be caused if air leaked from outside into the hallway?

7. Where should air from a chemical closet or similar area be vented?

Activity: Filtration

A diaphragm gauge is attached to an air handler system to show the pressure drop across a filter.

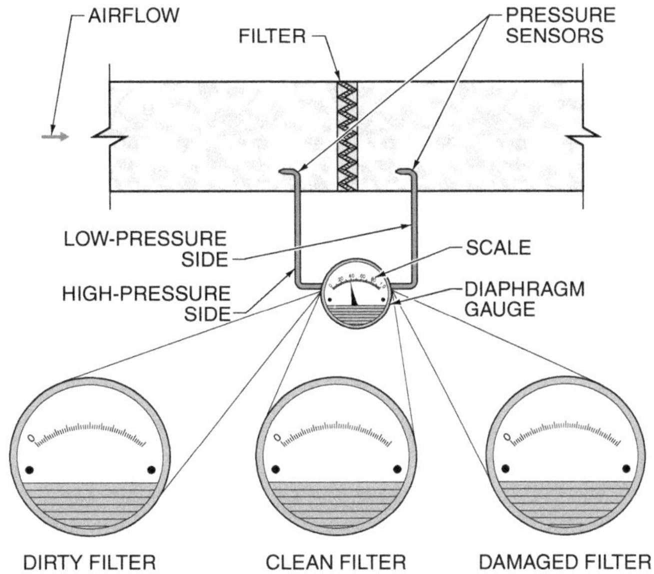

1. Sketch on the gauge the approximate needle position when the filter is very dirty and plugged with dirt particles.

2. Sketch on the gauge the approximate needle position when the filter is clean.

3. Sketch on the gauge the approximate needle position if the filter is damaged and air is blowing around the filter.

4. What does increasing efficiency mean in terms of the particle size a filter will trap?

5. What might happen to airflow if the air filters were changed from low-efficiency to high-efficiency filters?

6. What might happen to the supply air fan if the air filters were changed from low-efficiency to high-efficiency filters?

7. Why might a filter be changed long before it becomes clogged?

Activity: Control Systems

1. How is a closed-loop system different from an open-loop system?

2. List the important qualities of compressed air used in pneumatic control systems.

3. Describe two ways a DDC system can save energy.

4. What maintenance does a DDC system require?

Activity: Valve Signals

A normally open valve operator has an open to closed signal range of 4 psi to 8 psi.

1. What position will the valve be in when the air pressure signal is 2 psi?

2. What position will the valve be in when the air pressure signal is 4 psi?

3. What position will the valve be in when the air pressure signal is 6 psi?

4. What position will the valve be in when the air pressure signal is 8 psi?

5. What position will the valve be in when the air pressure signal is 10 psi?

Activity: Three-Way Valves

A three-way valve controls water flow through or around a cooling coil. The valve is currently actuated to provide maximum cooling.

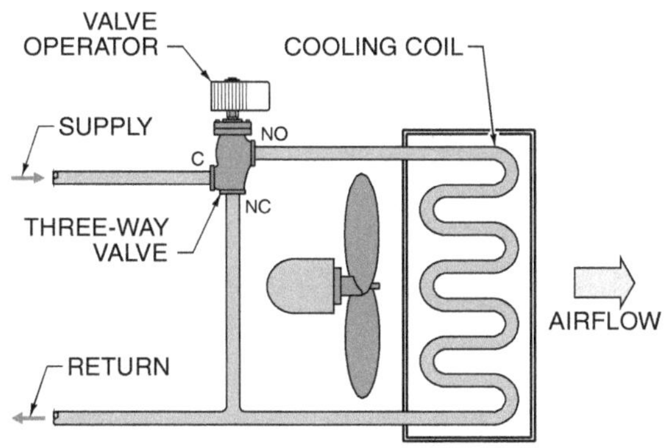

1. Indicate the direction of chilled water flow with arrows.

The three-way valve is set to provide no cooling.

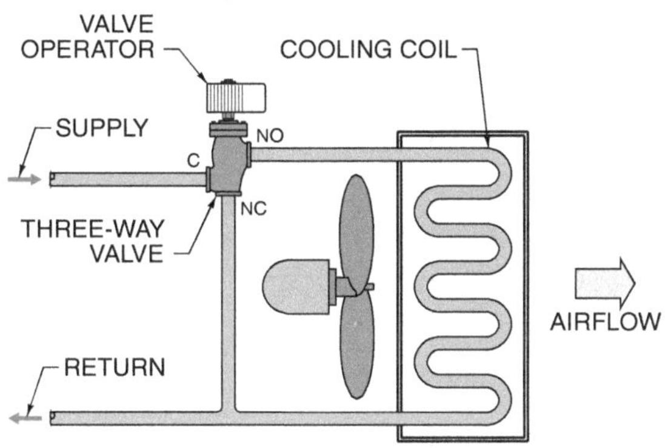

2. Indicate the direction of chilled water flow with arrows.

3. If the control signal were removed from the three-way valve, where would the water flow?

4. If the fan failed, how would water flow be affected?

Activity: HVAC System Troubleshooting

An office worker complains to the maintenance technician that the office space is too hot.

INVESTIGATE

Observed symptoms:

Outdoor temperature: 98°F

Relative humidity: 80%

Airflow to office: 220 CFM

Office temperature: 78°F

Cooling setpoint: 73°F

ISOLATE

Most likely causes of comfort problem:

- HVAC system cannot maintain temperature because it is extremely hot outside with high relative humidity
- poor airflow
- radiant heat from windows or hot wall

1. What steps should be taken to solve this problem?

The system is working within specifications, but it is running most of the time. It is simply too hot outside for the system to maintain the setpoint temperature.

2. What should be done next?

REMEDY

Discuss remedies to this problem with the instructor, fellow students, or fellow workers.

DOCUMENTATION

3. What documentation might need to be filled out?

4. Why is clear and complete documentation important in this situation?

REVIEW QUESTIONS

Name ______________________ **Date** ______________

True-False

T F **1.** A piston rod retracts when hydraulic oil flows to the cap end.

T F **2.** In a hydraulic jack, a large amount of pressure is used to create a large force.

T F **3.** If area is constant, force increases as pressure increases.

Multiple Choice

______________ **1.** A fluid pressure of 500 psi is exerted on a cylinder piston with and an area of 20 sq in. to produce ___ lb of force.

A. 1000
B. 5000
C. 10,000
D. 50,000

______________ **2.** ___ states that energy can be neither created nor destroyed but can be changed from one form to another.

A. Boyle's law
B. Newton's first law of motion
C. Newton's third law of motion
D. The law of conservation of energy

______________ **3.** A(n) ___ is a device that accepts a control signal and causes a mechanical motion.

A. actuator
B. pump
C. piston
D. directional-control valve

Completion

______________ **1.** A(n) ___ system is a fluid power system that transmits energy in an enclosed space using a liquid (hydraulic oil) under pressure.

______________ **2.** A(n) ___ is a container for storing fluid under little or no pressure.

______________ **3.** All hydraulic systems consist of a(n) ___, pathways, and one or more loads.

HYDRAULIC SYSTEM COMPONENTS

SECTION **11.2**

REVIEW QUESTIONS

Name ______________________________ **Date** ____________________

True-False

T F **1.** An axial reciprocating pump is a pump that consists of pistons located perpendicular to the pump shaft.

T F **2.** Hydraulic hose is less susceptible to damage from vibration or movement than rigid pipe.

T F **3.** The smaller the micron number of a filter material, the poorer the filtering capability.

T F **4.** Hydraulic motor speed is controlled by the rate of oil flow.

T F **5.** ISO 4406, *Hydraulic Fluid Power — Fluids — Method for Coding the Level of Contamination by Solid Particles,* uses a three-number code to determine the cleanliness of oil.

Multiple Choice

__________________ **1.** ___ is allowing a pump to run against little or no pressure.

A. Downsizing
B. Unloading
C. Depressurizing
D. Cushioning

__________________ **2.** A(n) ___ is a container in which fluid is stored under pressure.

A. basin
B. accumulator
C. well
D. reservoir

__________________ **3.** A ___ pump is a pump that has two inlets and outlets at opposite sides of the pump.

A. vane
B. gerotor
C. positive-displacement
D. balanced

__________________ **4.** ___ is the removal of the cover from a hydraulic hose to expose the reinforcement to attach a fitting.

A. Skiving
B. Skinning
C. Undressing
D. Stripping

______________ **5.** A ___ valve is a valve that directs or prevents flow through selected passages.
A. flow-control
B. pressure-sequence
C. directional-control
D. check

Completion

______________ **1.** A(n) ___ is an angled plate in contact with piston heads that moves the pistons in the cylinders of a reciprocating pump.

______________ **2.** ___ is the volume of fluid moved during each revolution, stroke, or cycle of a pump.

______________ **3.** A(n) ___ is a device that controls the pressure, direction, or rate of fluid flow.

______________ **4.** A hose ___ is a fitting that allows a line to be connected or disconnected by hand.

______________ **5.** A(n) ___ is a component that converts fluid pressure into linear mechanical force.

FLUID POWER SYSTEM PRINTREADING

SECTION 11.3

REVIEW QUESTIONS

Name ______________________________ **Date** ______________

True-False

T F **1.** A disadvantage of pictorial diagrams is that they are not standardized.

T F **2.** Cutaway diagrams are large and may span multiple pages.

T F **3.** An intersection with a dot means that the lines do not connect.

T F **4.** Most directional-control valves used in commercial and industrial applications are two-way, three-way, or four-way valves.

Multiple Choice

______________ **1.** In a schematic diagram, filled (all-black) triangles are used for ___ systems.
- A. hydraulic
- B. pneumatic
- C. electric
- D. steam

______________ **2.** On a standard cutaway diagram, the color ___ indicates intake flow to a pump drain line.
- A. yellow
- B. green
- C. blue
- D. red

______________ **3.** A(n) ___ line represents a main working line (pressure flow).
- A. dashed
- B. dotted
- C. solid
- D. arced

______________ **4.** A(n) ___ port receives fluid flow from the pump.
- A. A and B
- B. P
- C. T
- D. R

Completion

______________ **1.** A(n) ___ diagram is a diagram that shows the internal details of components and the path of fluid flow.

_______________ **2.** Typically, the most useful fluid power diagram is a(n) ___ diagram.

_______________ **3.** A(n) ___ is a certain arrangement that the spool of a directional-control valve can move into.

_______________ **4.** Actuators that are combined in-line with the valve symbol represent ___ logic.

HYDRAULIC SYSTEM MAINTENANCE

SECTION **11.4**

REVIEW QUESTIONS

Name ______________________________ **Date** ____________________

True-False

T F **1.** Accumulators should be charged with compressed air.

T F **2.** All hydraulic oils contain additives such as foaming and rust inhibitors.

T F **3.** Hoses should be stored by hanging them on nails or hooks in a closet.

Multiple Choice

______________ **1.** Hardness, brittleness, or spongy consistency of the seal material indicates damage from ___.
- A. excessive temperatures
- B. improper cleaning
- C. excessive force
- D. incorrect parts

______________ **2.** Hydraulic oil that is too thick can cause ___.
- A. internal leakage
- B. hose cracking or splitting
- C. sluggish operation
- D. overheating

______________ **3.** Hoses should be stored ___.
- A. in warm, dry areas
- B. in cool, dry areas
- C. in direct sunlight
- D. on hooks

Completion

______________ **1.** Cold hydraulic oil can cause problems due to ___ viscosity, which results in poor lubrication.

______________ **2.** ___ are required to support long hose runs or to keep the hose away from moving parts.

______________ **3.** The inside of piping or tubing should be kept as smooth and clean as possible, and piping runs should be kept as ___ as possible.

HYDRAULIC SYSTEM TROUBLESHOOTING

SECTION 11.5

REVIEW QUESTIONS

Name ______________________________ Date ____________________

True-False

T F **1.** Leaks reduce the capacity of equipment to perform work.

T F **2.** Internal leaks generate heat in hydraulic oil, causing thickening.

T F **3.** Head pressure loss can be caused by a long pipe run, rough internal surfaces of the pipe, or fast fluid flow.

Multiple Choice

______________ **1.** If an actuator is not producing enough force, the problem is likely related to ___.

A. wear and corrosion of the actuator
B. dirty system components
C. improper lubrication
D. lack of pressure in the system

______________ **2.** ___ are used to measure motor speed.

A. Hygrometers
B. Hydraulic testers
C. Tachometers
D. Unloaders

______________ **3.** ___ is a process in which microscopic gas bubbles expand in a vacuum and suddenly implode when entering a pressurized area.

A. Cavitation
B. Implosion
C. Foaming
D. Popping

Completion

______________ **1.** A hydraulic piston rod is controlled by the position of a(n) ___ valve, which is operated by an electric solenoid or manual switches.

______________ **2.** A(n) ___ is a device that measures the pressure, flow, and temperature of hydraulic oil in a system.

______________ **3.** As fluid flows through pipes, some energy generated at a pump is converted into heat due to ___.

PNEUMATIC SYSTEMS

SECTION 11.6

REVIEW QUESTIONS

Name ______________________________ **Date** ____________________

True-False

T F **1.** The physical characteristics of gases are affected by pressure, volume, and temperature.

T F **2.** More air can be stored and used in a system if it is heated before being compressed.

T F **3.** The most common method for regulating pressure in a receiver for large compressors is by starting and stopping the compressor.

T F **4.** Pneumatic tools present no danger of electrical shock.

Multiple Choice

______________ **1.** Pneumatic systems usually operate at lower ___ than hydraulic systems.

A. capacities
B. pressures
C. efficiencies
D. volumes

______________ **2.** A(n) ___ is an air tank that stores compressed air and allows it to cool before use.

A. receiver
B. intensifier
C. intercooler
D. unloader

______________ **3.** A(n) ___ is a device that allows a compressor to operate without adding pressure to the receiver.

A. intensifier
B. relief valve
C. receiver
D. unloader

______________ **4.** Air at 100 psi applied to a piston with a surface area of 5 sq in. exerts ___ lb of force in a straight line through the ram to its surface, pressurizing the oil in a small cylinder.

A. 250
B. 500
C. 1000
D. 5000

Completion

______________________ **1.** The ___ are the predictable relationships between the volume, pressure, and temperature of a gas.

______________________ **2.** Cleaning operations require air pressure below ___ psi.

______________________ **3.** A(n) ___ is a device that converts low-pressure fluid power into high-pressure fluid power.

______________________ **4.** A(n) ___ system is a fluid power system that transmits energy using a gas (typically compressed air).

PNEUMATIC SYSTEM COMPONENTS

SECTION **11.7**

REVIEW QUESTIONS

Name ______________________________ **Date** ____________________

True-False

T F **1.** A pressure-relief valve and a high-pressure cutout are located on the outlet lines of all air compressors.

T F **2.** Pneumatic system pipe must be plumbed with a downward slope so that water in the lines flows into drains.

T F **3.** The warmer the air drawn into a compressor, the more efficiently the compressor operates.

Multiple Choice

______________ **1.** A ___ compressor is a compressor that uses two or three cylinders, each with progressively smaller diameters, to produce progressively higher pressures.
- A. screw
- B. multistage
- C. vane
- D. multicylinder

______________ **2.** A(n) ___ is a shell-and-tube heat exchanger that uses water to cool compressed air.
- A. unloader
- B. refrigerant dryer
- C. aftercooler
- D. intercooler

______________ **3.** A ___ is a drying agent that removes water vapor by absorption.
- A. condenser
- B. compressor
- C. refrigerant dryer
- D. desiccant

Completion

______________ **1.** An air ___ is a component that takes air from the atmosphere and compresses it to increase its pressure.

______________ **2.** An oil and moisture ___ is a device that removes oil and water droplets from compressed air by forcing it to change direction quickly.

______________ **3.** A(n) ___ is a component that injects a mist of oil into a compressed air line for lubrication of pneumatic tools and internal motor parts.

PNEUMATIC SYSTEM MAINTENANCE

SECTION **11.8**

REVIEW QUESTIONS

Name ______________________________ **Date** ____________________

True-False

T F **1.** If possible, line pressure should be shut off to stop a broken air line from whipping around while exhausting compressed air.

T F **2.** System piping should be opened without first bleeding off air.

T F **3.** Excessive coats of paint to any part of a pneumatic system can increase insulation, which prevents proper heat transfer.

Multiple Choice

______________ **1.** Too much compressor lubricant can cause ___.
A. excessive heat
B. leaks
C. premature wear on components
D. a component to clog

______________ **2.** Compressors should always be ___ when started.
A. under load
B. unloaded
C. room temperature
D. disconnected

______________ **3.** When a compressor is at or near full speed and the bleed-off valve closes, air is then sent through the ___ to supply compressed air to the system.
A. check valve
B. butterfly valve
C. filter-dryer
D. drain

Completion

______________ **1.** A(n) ___ valve is a valve that allows the pressure against the discharge valves to be released at startup.

______________ **2.** ___ that are intermittently used or held for a long time in one position should be operated occasionally to ensure free movement.

______________ **3.** Ensuring a supply of clean, ___ air is the most important task in pneumatic system maintenance.

PNEUMATIC SYSTEM TROUBLESHOOTING

SECTION 11.9

REVIEW QUESTIONS

Name ______________________________ Date ____________________

True-False

T F 1. The most common problem in pneumatic systems is excessive air pressure or airflow caused by disconnected, kinked, or worn lines.

T F 2. A common cause of pneumatic system problems is compressor controls set at incorrect pressure settings.

T F 3. If an actuator does not operate with the load disconnected, the load may be too great.

Multiple Choice

______________ **1.** When checking for pressure or flow problems, technicians should work ___.
A. backward from the compressor
B. backward from the actuators
C. forward from the shutoff valve
D. forward from the filter

______________ **2.** To test an actuator, the first step is to ___.
A. disconnect the load
B. energize the actuator
C. increase the air pressure
D. disconnect the air line

______________ **3.** ___ commonly occur at connection points in a pneumatic system.
A. Condensation droplets
B. Dirty areas
C. Hot spots
D. Leaks

Completion

______________ **1.** The application of soapy water produces ___ when applied to a leak.

______________ **2.** A(n) ___ is a component on an actuator that applies tension to a seal or packing to prevent leakage around the shaft.

______________ **3.** Dirt and debris can block small ___ passages and cause operating problems.

FLUID POWER SYSTEMS

Chapter 11

ACTIVITIES

Name ______________________________ Date ____________________

Hydraulic Force

_______________ 1. A pressure of ___ psi is required for A.

_______________ 2. The area of the piston required for B is ___ sq in.

_______________ 3. A force of ___ lb is produced for C.

_______________ 4. A force of ___ lb is produced for D.

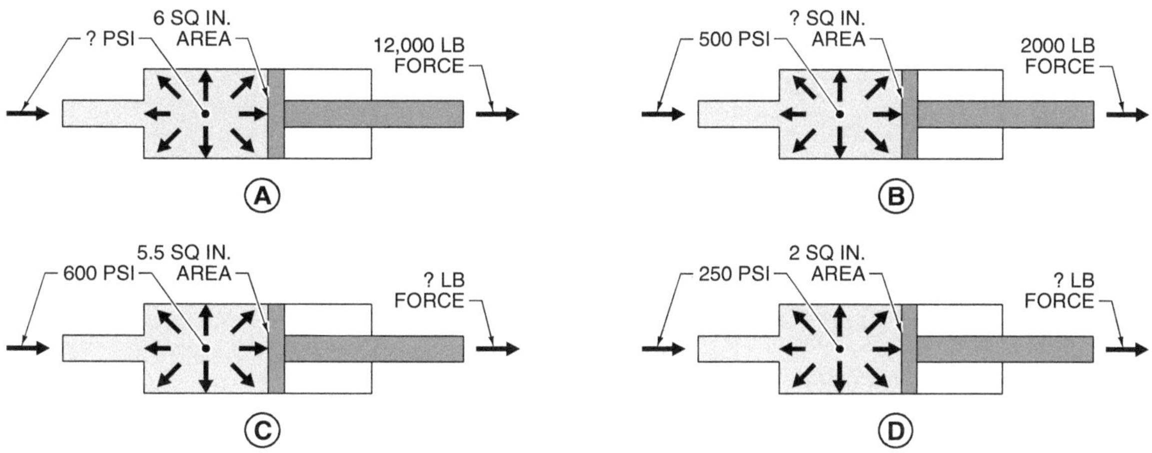

Vane Pumps

_______________ 1. Housing

_______________ 2. Shaft

_______________ 3. Rotor

_______________ 4. Inlet

_______________ 5. Outlet

_______________ 6. Seal

_______________ 7. Movable vanes

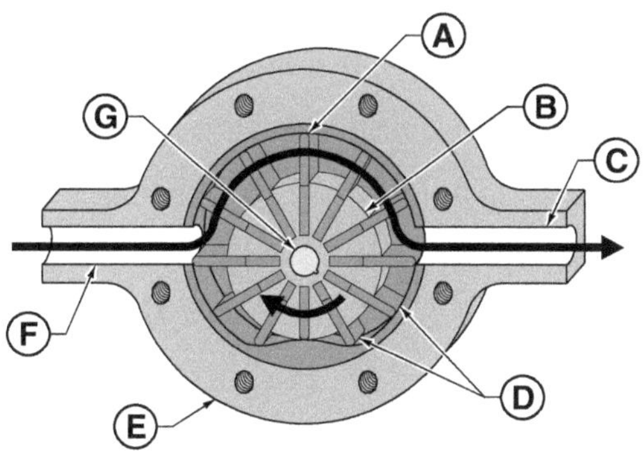

Gear Pumps

__________ **1.** Gerotor

__________ **2.** External

__________ **3.** Internal

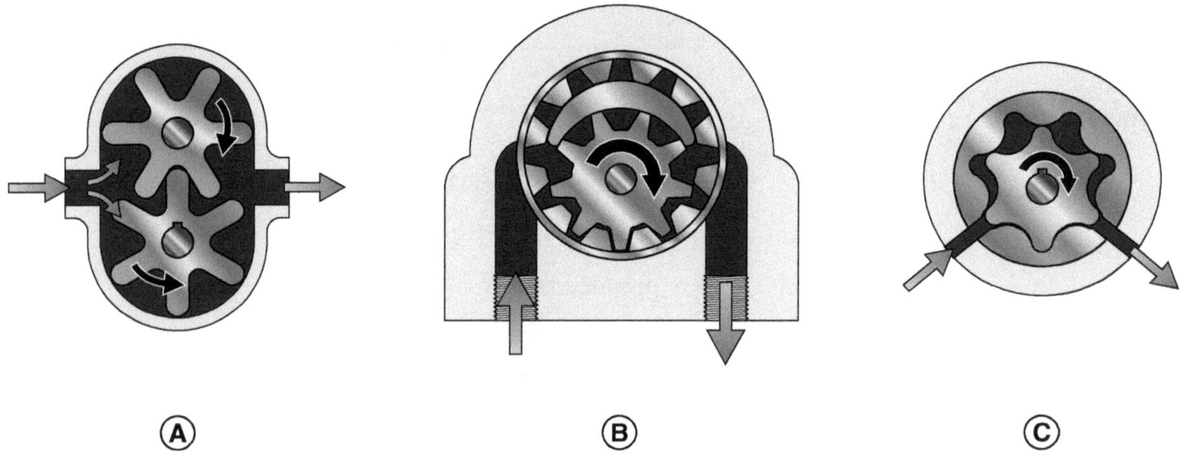

Hydraulic Hose Coupling

__________ **1.** Seal

__________ **2.** Spring

__________ **3.** Sleeve

__________ **4.** Poppet valve

__________ **5.** Poppet seal

__________ **6.** Valve stop

__________ **7.** Coupling threads

__________ **8.** Ball locking mechanism

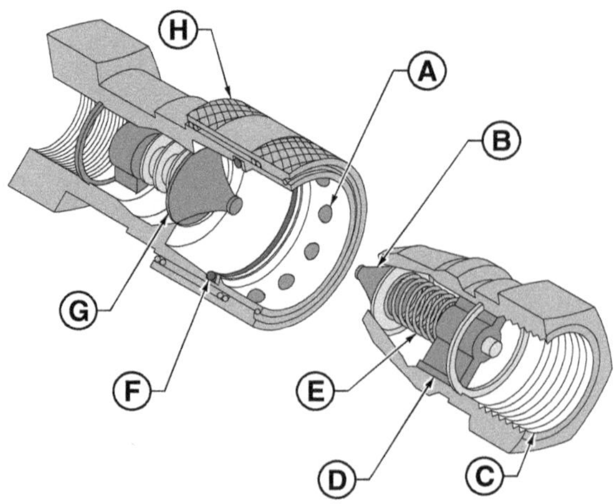

Pneumatic Systems

_______________ **1.** Air receiver

_______________ **2.** Shutoff valve

_______________ **3.** Lubricator

_______________ **4.** Regulator

_______________ **5.** Filter

_______________ **6.** Pressure-relief valve

_______________ **7.** Compressor

_______________ **8.** Inlet air cleaner

_______________ **9.** Electrical control panel

_______________ **10.** Automatic drains

_______________ **11.** Refrigerant dryer

_______________ **12.** Aftercooler

_______________ **13.** Pressure gauge

_______________ **14.** Oil and moisture separator

_______________ **15.** Intercooler

_______________ **16.** Main line

_______________ **17.** High-pressure cutout and compressor operating controls

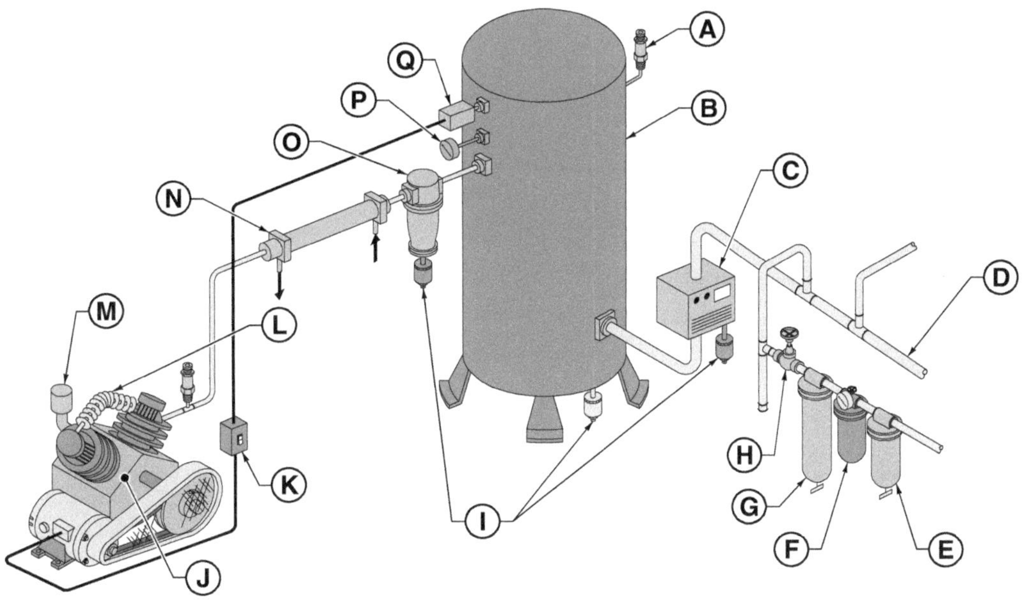

Filter, Regulator, and Lubricator (FRL) Stations

__________ **1.** Air-powered tool

__________ **2.** Quick disconnect

__________ **3.** Regulator

__________ **4.** Filter

__________ **5.** Shutoff valve

__________ **6.** Lubricator

__________ **7.** Flexible air line

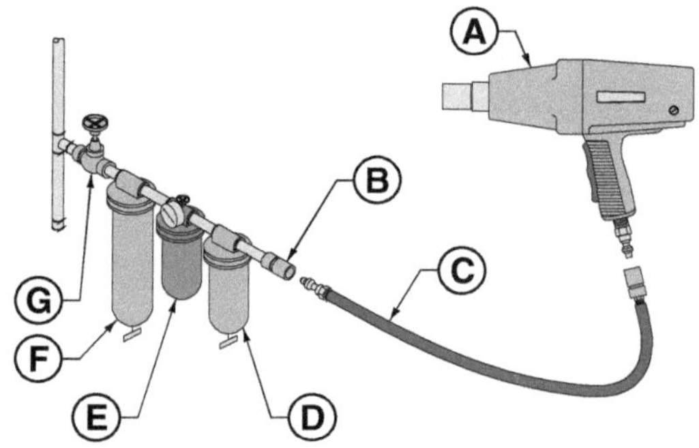

Hydraulic System Components — Pictorial

__________ **1.** Reservoir

__________ **2.** Filter

__________ **3.** Pump

__________ **4.** Return line

__________ **5.** Directional-control valve

__________ **6.** Pressure-relief valve

__________ **7.** Pressure gauge

__________ **8.** Piston rod

__________ **9.** Piston

__________ **10.** Cylinder

__________ **11.** Flow-control valve

__________ **12.** Check valve

__________ **13.** Electric motor

__________ **14.** Rod end

__________ **15.** Cap end

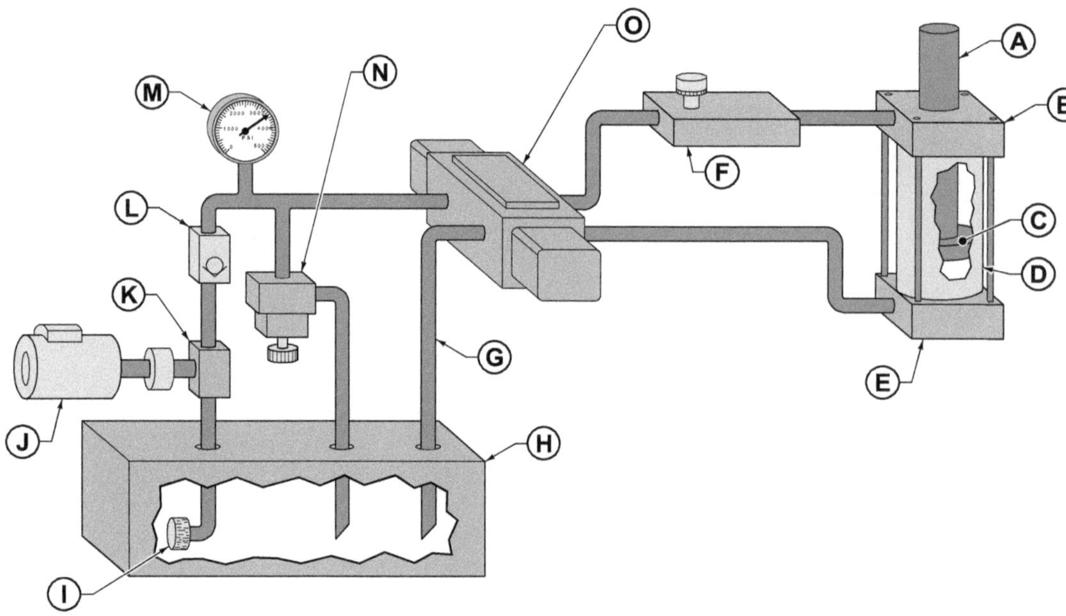

Hydraulic System Components — Schematic

____________________	**1.** Piston rod	____________________	**8.** Cylinder
____________________	**2.** Cap end	____________________	**9.** Reservoir
____________________	**3.** Rod end	____________________	**10.** Filter
____________________	**4.** Flow-control valve	____________________	**11.** Pump
____________________	**5.** Directional-control valve	____________________	**12.** Pressure gauge
____________________	**6.** Pressure-relief valve	____________________	**13.** Return line
____________________	**7.** Check valve	____________________	**14.** Electric motor

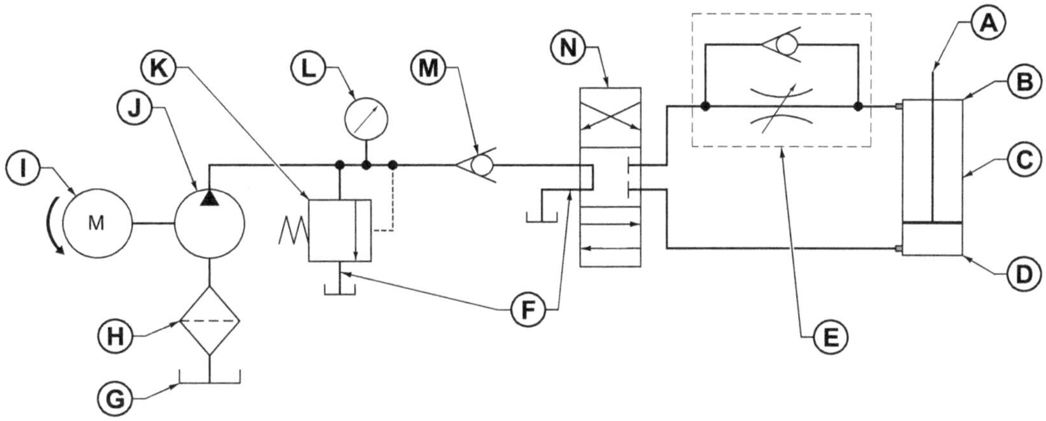

Activity: Pump Operation

An external gear pump is used to circulate oil through a hydraulic system.

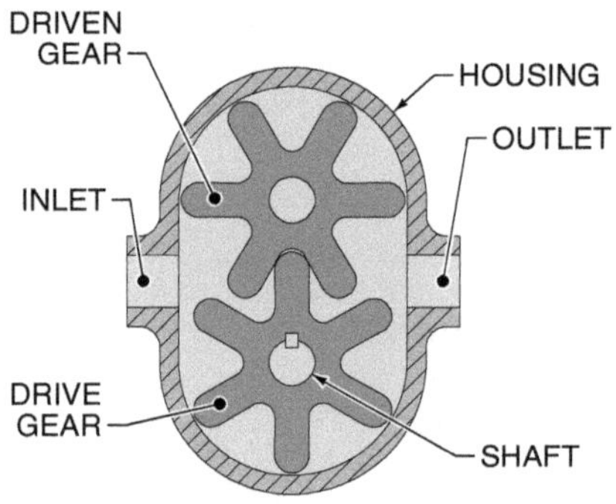

1. Label the high- and low-pressure areas.

2. Sketch the direction of oil flow into, out of, and through the pump.

3. Indicate and label two places where the rotor seals with the pump housing.

4. What problem will occur if oil flow were restricted at the pump inlet?

5. What would the outlet oil look like if there were a leak on the inlet side of the pump?

6. How would a leak in the inlet side of the pump be located?

7. What problem might occur if the oil were extremely cold?

8. Does a pump develop flow or pressure?

Activity: Pressure and Flow-Control Valves

A pressure-relief valve is a valve that sets a maximum operating pressure level for a circuit to protect the circuit from overpressure.

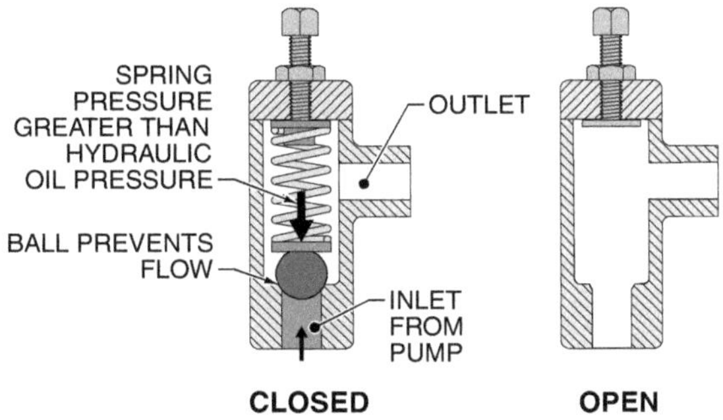

1. Sketch the position of the ball and spring and the direction of oil flow in the pressure-relief valve when it is open.

A pressure-reducing valve is a valve that limits the maximum pressure at its outlet, regardless of the inlet pressure.

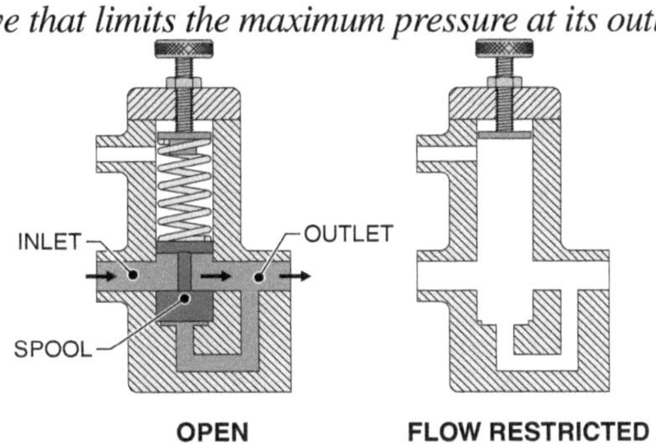

2. Sketch the position of the spool and spring and the direction of oil flow in the pressure-reducing valve when it is restricting flow.

A pressure-sequence valve is a pressure-operated valve that diverts flow to a secondary actuator while holding pressure on the primary actuator at a predetermined minimum value after the primary actuator completes its travel.

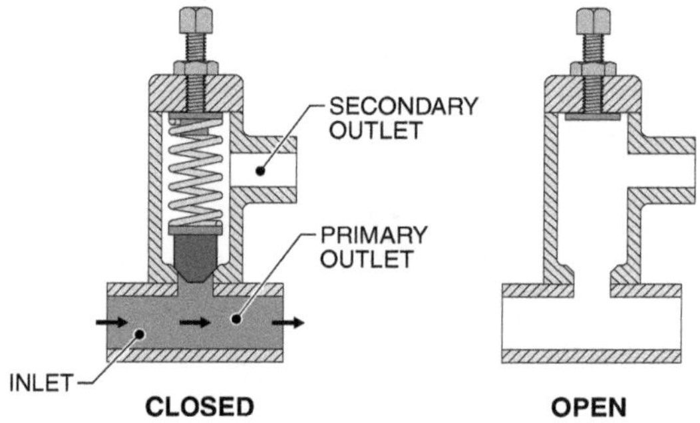

3. Sketch the position of the spool and spring and the direction of oil flow in the pressure-sequence valve when it is directing oil to the secondary outlet.

An unloading valve is a pressure-control valve that directs hydraulic oil from the pump to the reservoir after system pressure has been reached.

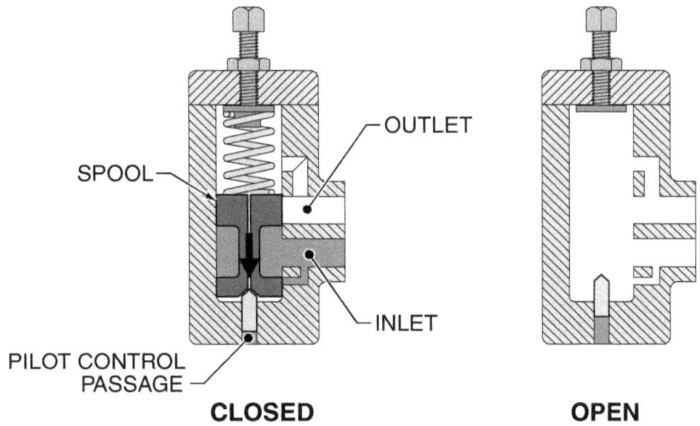

4. Sketch the position of the spool and spring and the direction of oil flow in the unloading valve when it is open.

5. Why should the spring tension not be adjusted when pressure or flow measurement equipment is not attached to the system?

6. Why must system specifications be consulted when adjusting pressure or flow?

7. What two dangerous situations could occur if pressure and flow were altered carelessly?

8. In which two places must changes to pressure and flow be noted?

9. What part of an unloading valve is most affected by dirty oil?

10. What long-term problem could develop if the hydraulic oil were not sealing the metal parts of these valves properly?

Activity: Unloading Valves

A hydraulic system includes an unloading valve to direct oil back to the reservoir when a specific pressure has been reached. The unloading valve is currently closed.

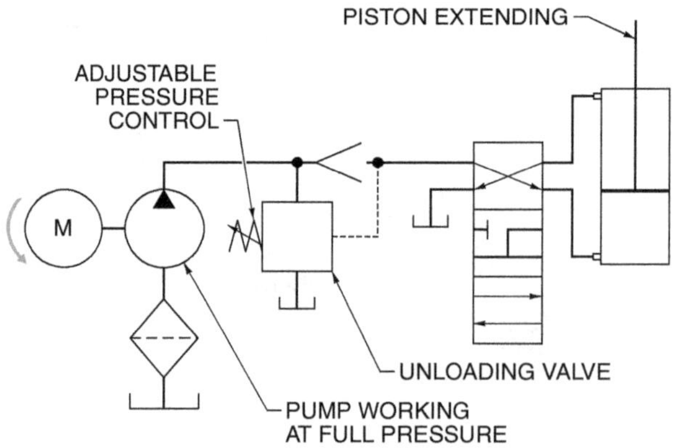

1. Sketch the check valve position, the unloading valve position, and oil flow directions.

System pressure reaches a certain level and the unloading valve opens.

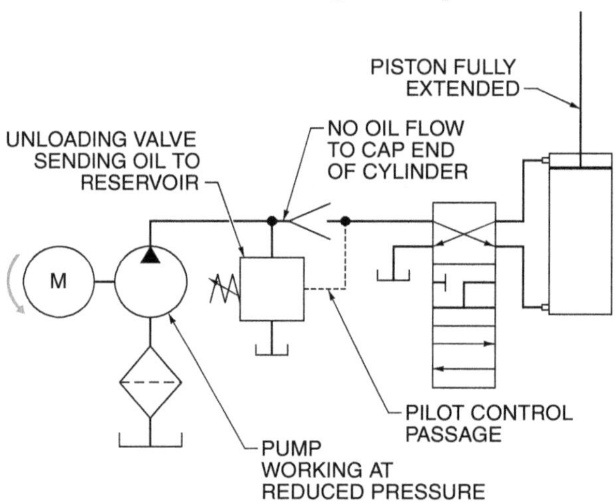

2. Sketch the check valve position, the unloading valve position, and oil flow directions.

3. What is a likely cause of a system with an unloading valve not supplying enough pressure to the loads?

4. What is a likely cause of a system with an unloading valve not unloading?

5. What damage could occur if the unloading valve does not unload at its set pressure?

Activity: Directional-Control Valves

A directional-control valve is used in a hydraulic piston control circuit.

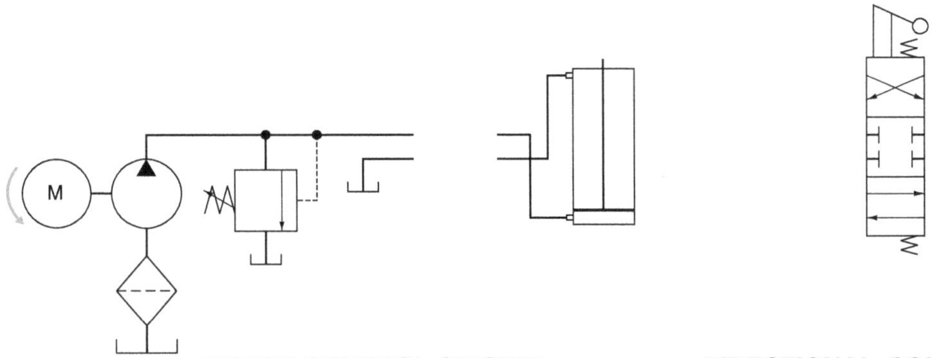

PISTON CONTROL CIRCUIT **DIRECTIONAL–CONTROL VALVE**

1. Sketch the directional-control valve in the correct position to extend the piston rod.

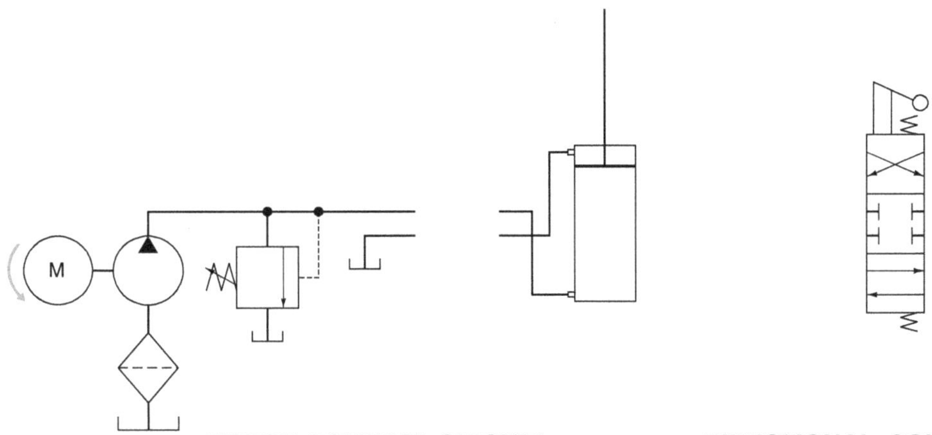

PISTON CONTROL CIRCUIT **DIRECTIONAL–CONTROL VALVE**

2. Sketch the directional-control valve in the correct position to retract the piston rod.

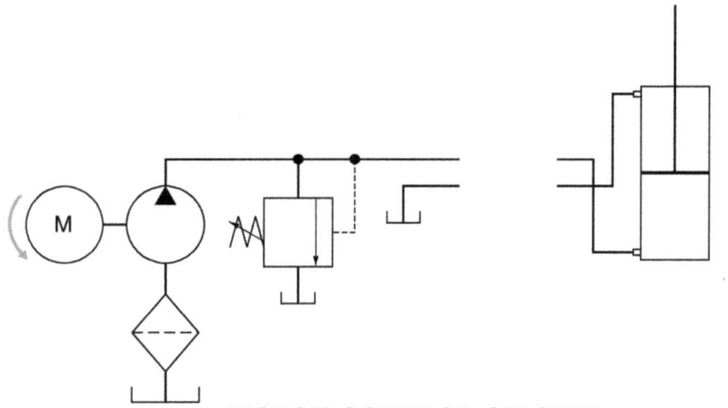

PISTON CONTROL CIRCUIT

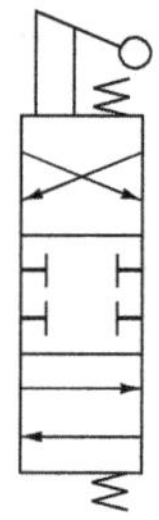

DIRECTIONAL–CONTROL VALVE

3. Sketch the directional-control valve in the neutral position.

4. List two problems that could cause the piston to slowly retract in the neutral position.

5. Is the piston single-acting or double-acting?

6. Is the piston balanced or unbalanced?

7. If the piston becomes damaged by slamming hard as it retracts and extends, what type of cylinder could be substituted to prevent this damage?

8. What maintenance does this hydraulic system require?

Activity: Hydraulic Power Circuit

A hydraulic power circuit uses two directional-control valves, each actuated by solenoids controlled by a PLC. One valve controls a hydraulic cylinder, and the other controls a rotary actuator.

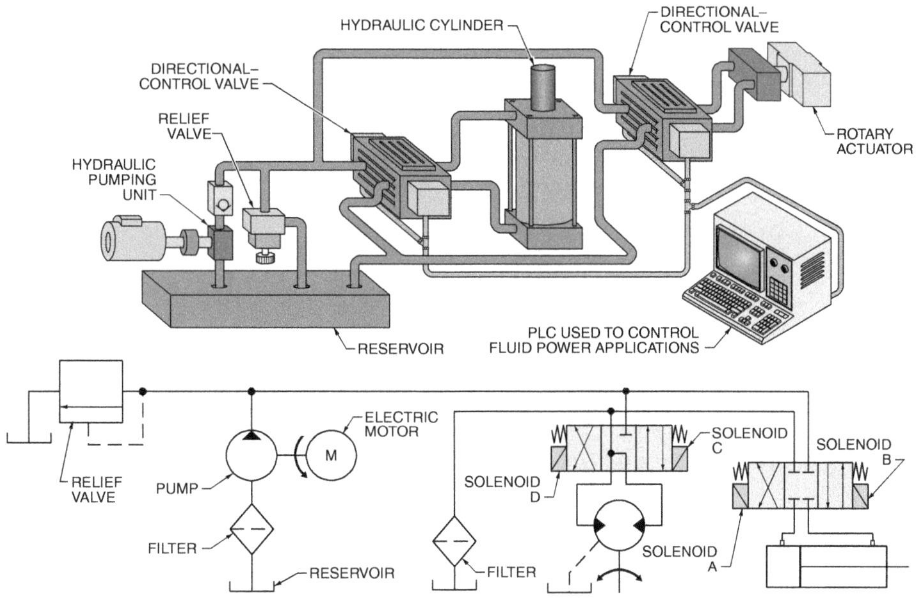

1. If solenoid A overheated and failed, which hydraulic device would be affected?

2. What would happen to system operation if the relief valve setting were too low?

3. What mechanical problem could prevent the cylinder from extending?

4. If power were lost to the PLC, would the hydraulic system continue to operate?

5. What predictive maintenance practice would be useful in maintaining the quality of the hydraulic oil?

6. What problems could result in the rest of the system if the hydraulic filter becomes clogged and the oil bypasses the filter?

7. What problem could cause the cylinder to not stay in position when its directional-control valve is in the neutral position?

8. How could the PLC be used to help troubleshoot the system?

9. What would happen to oil flow if the pump seals and valves were becoming worn?

Activity: Pneumatic Systems

Answer the following questions about pneumatic systems and their components.

1. What is the purpose of the intercooler?

2. Is the operation of the high-pressure cutout and the operating control similar to controls on a boiler system?

3. Why must oil and water be removed from the compressed air?

4. How does a refrigerated dryer remove moisture from the air?

5. What dangerous situation can occur if a pressurized air hose breaks, and how should this situation be handled?

6. Why are pneumatic systems fitted with pressure-relief valves?

7. Should an oil lubricator be installed in an air system that supplies air to HVAC controls or medical tools?

8. Should a filter and pressure regulator be installed in an air system that supplies air to HVAC controls or medical tools?

9. What might happen to the temperature of the compressed air if it were being used so quickly that it flowed directly through the receiver rather than being held there for several hours before use?

FACILITY EFFICIENCIES

SECTION 12.1

REVIEW QUESTIONS

Name ______________________________ **Date** ____________________

True-False

T F **1.** Energy efficiency can be improved by either accomplishing more output with the same amount of energy input or maintaining the same output while requiring less energy input.

T F **2.** Reducing energy consumption can increase the utility bills of a facility.

T F **3.** Inventory waste is unproductive time due to needed parts or materials being unavailable.

T F **4.** Non-energy resources continue to cycle through the ecosystem and industry, but they become useless once they are polluted.

T F **5.** Carbon dioxide and other gases accumulate and trap heat within the Earth's atmosphere.

T F **6.** A carbon penalty is a regulatory action associated with a certain amount of carbon dioxide emissions.

T F **7.** Watering landscapes during warmer times of the day prevents excessive water evaporation.

T F **8.** Diagnosing and remedying a problem as quickly as possible minimizes costs and waste associated with downtime.

Multiple Choice

______________ **1.** Energy ___ is the measure of how much work is done with a certain amount of energy input.
- A. conservation
- B. efficiency
- C. productivity
- D. input

______________ **2.** ___ is waste due to unnecessary steps performed on each product or task.
- A. Overproduction waste
- B. Inventory waste
- C. Waiting
- D. Extra processing

______ **3.** ___ is the arrangement of all necessary items in an area so that they can be easily found and selected for use.
A. Sorting
B. Shining
C. Setting in order
D. Standardizing

______ **4.** ___ is a requirement for ongoing organization and cleanliness.
A. Sustaining
B. Sorting
C. Setting in order
D. Shining

______ **5.** ___ energy is energy from a source that is continuously replenished.
A. Reused
B. Reclaimed
C. Recycled
D. Renewable

______ **6.** The most common renewable energy source currently installed on commercial and industrial properties is ___.
A. solar panels
B. wind turbines
C. geothermal power plants
D. waterwheels

______ **7.** ___ appliances are certified as water efficient by the Environmental Protection Agency (EPA).
A. Energy Star®
B. WaterSense™
C. Low-flow
D. Drought-resistant

______ **8.** The ___ green building rating system from the U.S. Green Building Council is the nationally accepted standard for the design, construction, and operation of green buildings.
A. WaterSense™ efficiency
B. EnergyStar®
C. Leadership in Energy and Environmental Design (LEED)
D. Occupational Information Network (O*NET)

Completion

______ **1.** Energy ___ is a reduction in energy use through changes in consumer behavior.

______ **2.** Reducing ___ is a key goal in energy efficiency and energy conservation practices.

______ **3.** The five categories of organization are sorting, setting in order, shining, standardizing, and ___.

______ **4.** ___ is the selection and efficient use of resources in order to prevent their depletion.

ENERGY AUDITING

SECTION **12.2**

REVIEW QUESTIONS

Name ______________________________ **Date** ____________________

True-False

T F **1.** An energy audit is an effective survey of only electricity consumption.

T F **2.** Energy auditing ranges from a quick walk-through to a comprehensive analysis of all energy consumption.

T F **3.** Typically, only major problems, such as leaks or inappropriate temperature setpoints, are uncovered during a general energy audit.

Multiple Choice

______________ **1.** During an audit, detailed information about energy use is collected, and strategies to reduce the consumption are prioritized in order of ___.
- A. level of intervention required
- B. cost effectiveness
- C. energy conserved
- D. length of estimated intervention

______________ **2.** A general energy audit requires a record of ___ from the past 12 to 36 months, which reveals short- and long-term consumption patterns and rate changes.
- A. inventory purchases
- B. troubleshooting reports
- C. utility bills
- D. production targets

______________ **3.** Budget decisions are based on the expected ___ that applies to all expenditures, energy related or not.
- A. energy cost
- B. labor time
- C. cost of materials and labor
- D. return on investment

Completion

______________ **1.** A(n) ___ is a comprehensive review of the energy use of a facility and a report on ways to reduce energy use through changes to buildings, equipment, and procedures.

_______________ **2.** The three common types of audits include preliminary, general, and ___ grade.

_______________ **3.** A(n) ___ energy audit is an overview of the major energy-consuming processes in a facility to identify only significant inefficiencies.

PLAN IMPLEMENTATION

SECTION **12.3**

REVIEW QUESTIONS

Name ______________________________ **Date** ____________________

True-False

T F **1.** The cost of the electricity to power an electric motor is likely most of the total life-cycle cost.

T F **2.** Utilities may provide financial incentives for their customers to make smarter use of available energy.

T F **3.** An investment with a return on investment (ROI) of five years will lose the savings from that investment after the first five years.

Multiple Choice

____________ **1.** One of the best ways to change reluctant attitudes toward energy efficiency improvements is to start with ___.

A. thorough research into the cost analysis
B. low- or no-cost projects
C. major projects
D. facility-wide changes to procedures

____________ **2.** The International Standards Organization (ISO) 50001, *Energy Management Systems* (EnMS), is aimed at improving energy efficiency in ___ facilities.

A. residential
B. schools and public
C. manufacturing and commercial
D. government

____________ **3.** A one-time investment of $100 that yields a savings of $20 per year has an ROI of ___.

A. 10%
B. 80%
C. 5 years
D. 10 years

Completion

____________ **1.** A(n) ___ is a financial comparison of an investment to the resulting economic benefits.

____________________ **2.** A return on investment (ROI) is commonly represented as a(n) ___ of savings or as the amount of time for energy savings to pay for the investment.

____________________ **3.** A(n) ___ is a return on investment (ROI) analysis that accounts for all the expenses and savings for an investment over a period of many years, usually 20 or 30 years.

SYSTEM EFFICIENCY MEASURES

SECTION **12.4**

REVIEW QUESTIONS

Name ______________________________ **Date** ____________________

True-False

T F **1.** A common problem with bearing lubrication is undergreasing.

T F **2.** Vibration patterns or signatures in equipment may indicate gear wear or misalignment.

T F **3.** In general, a motor is at its best efficiency from about 50% to 75% load factor.

T F **4.** Variable-frequency drives (VFDs) use electronics to control the speed of a motor by controlling the frequency of the direct current (DC) supplied.

T F **5.** Deterioration of lighting fixtures and dust accumulation can reduce illumination by 50% or more.

T F **6.** It takes more energy to cool a space to 45°F than it does to cool that same space to 43°F.

T F **7.** The most effective way to save energy is to use energy only when needed.

T F **8.** Defrost cycles use up to 21% of the energy in a cold storage facility.

Multiple Choice

______________ **1.** Using a ___ on a storage container is the best method for transferring lubricant to a machine.

A. flexible hose
B. rigid hose
C. funnel
D. bucket

______________ **2.** ___ is the control of facility indoor lighting in a manner that maximizes use of natural sunlight from windows and skylights and uses artificial light only when necessary.

A. Unloading
B. Load shedding
C. Natural lighting
D. Daylighting

______________ **3.** Electric utilities commonly add a surcharge to electrical bills to compensate for the extra electricity they must generate to overcome ___.

A. undersized motors
B. poor power factor
C. poor luminous efficacy
D. variable-frequency drive (VFD) resistance

______________ **4.** ___ is an energy-conservation strategy in which controllers automatically turn off non-essential electrical systems during peak demand periods when electricity costs are higher.
A. Load shifting
B. Daylighting
C. Load shedding
D. Load cycling

______________ **5.** If switch contacts are too small for the current being carried, misaligned, or corroded or pitted due to heavy use, then they have unnecessarily ___.
A. low temperatures
B. low current draw
C. high resistance
D. high conductivity

______________ **6.** ___ occupancy sensors detect the radiant heat of a person.
A. Ultraviolet (UV)
B. Thermal imaging
C. Ultrasonic
D. Passive infrared (PIR)

______________ **7.** A heat exchanger installed in the stack just outside a boiler is called a(n) ___.
A. economizer
B. steam trap
C. steam drum
D. draft coil

______________ **8.** Cooler inlet air going into a compressor increases ___.
A. vibration problems
B. efficiency
C. harmonics
D. the likelihood of equipment failure

Completion

______________ **1.** ___ is the ratio of the average motor load to the rated motor load.

______________ **2.** ___ is the ratio of the light output (lumens) of a lamp to the electrical power input (watts).

______________ **3.** A(n) ___ load is a plugged-in load that continuously consumes a small amount of power whether or not the load is on.

______________ **4.** When a new building or a building system is about to come into service, it is ___.

______________ **5.** Load ___ is the rescheduling of predictable electrical loads to low-demand times, such as overnight.

______________ **6.** Scale buildup lowers ___ efficiency.

______________ **7.** Excessive pressure drop in an inactive section might indicate ___.

______________ **8.** ___ indicates the amount and types of particles being carried in oil.

EFFICIENCY AND SUSTAINABILITY

Chapter 12

ACTIVITIES

Name ______________________________ Date ____________________

Lean Principles

______ 1. Transportation

______ 2. Overproduction

______ 3. Waiting

______ 4. Inventory waste

______ 5. Defects

______ 6. Extra processing

______ 7. Movement waste

A. Inefficient motion of persons or material-handling equipment when performing a task

B. Work that is done beyond what is required

C. Unproductive time due to needed parts or materials being unavailable

D. Waste due to unnecessary steps performed on each product or task

E. Waste in which work must be repeated to correct or replace the product or task

F. Unnecessary moving of parts or materials during a process

G. Stocking of unnecessary parts or materials that consume space, contribute to disorganization in inventory stocks, and cause unnecessary spending

5S Organization

______ 1. Sorting

______ 2. Setting in order

______ 3. Shining

______ 4. Standardizing

______ 5. Sustaining

A. Creating easily repeatable processes to produce high-quality products

B. Eliminating unnecessary work

C. Keeping the workplace and all tools and equipment clean and neat

D. Arranging all necessary items in an area so they can be easily found and selected for use

E. Requiring ongoing organization and cleanliness

Energy Audit Process

______ **1.** Choose cost-effective plan

______ **2.** Measure energy use

______ **3.** Gather information

______ **4.** Develop reduction strategies

______ **5.** Implement plan

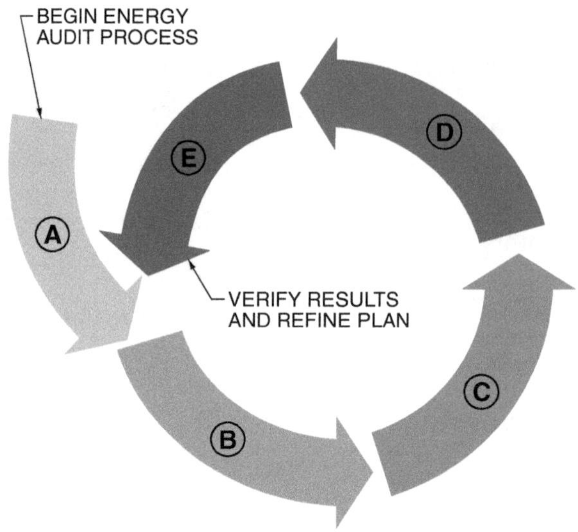

Return on Investment (ROI)

______ **1.** An investment of $100 yields a one-time savings of $200, for an ROI of ___%.

______ **2.** An investment of $100 yields an annual savings of $50, for an annual ROI of ___%.

______ **3.** For an annual ROI of 25%, an investment of $200 must yield $___ in savings each year.

______ **4.** For an investment of $500 with an annual ROI of 10%, the payback period is ___ years.

______ **5.** In order to pay back an investment of $400 within 4 years, the annual savings must be at least $___.

______ **6.** In order to pay back an investment of $800 within 10 years, the annual ROI must be at least ___%.

Activity: Efficiency, Conservation, and Sustainability

Answer the following questions about efficiency, conservation, and sustainability.

1. What are the two fundamental ways in which energy efficiency can be improved?

2. Are both methods equally viable for most facilities?

3. What is energy conservation?

4. How does energy efficiency also support sustainability?

5. What are some ways to improve energy source sustainability?

6. Do sustainability efforts apply only to energy sources?

7. How is sustainability related to preventing climate change?

Activity: Renewable Energy Credits (RECs)

Answer the following questions about Renewable Energy Credits (RECs).

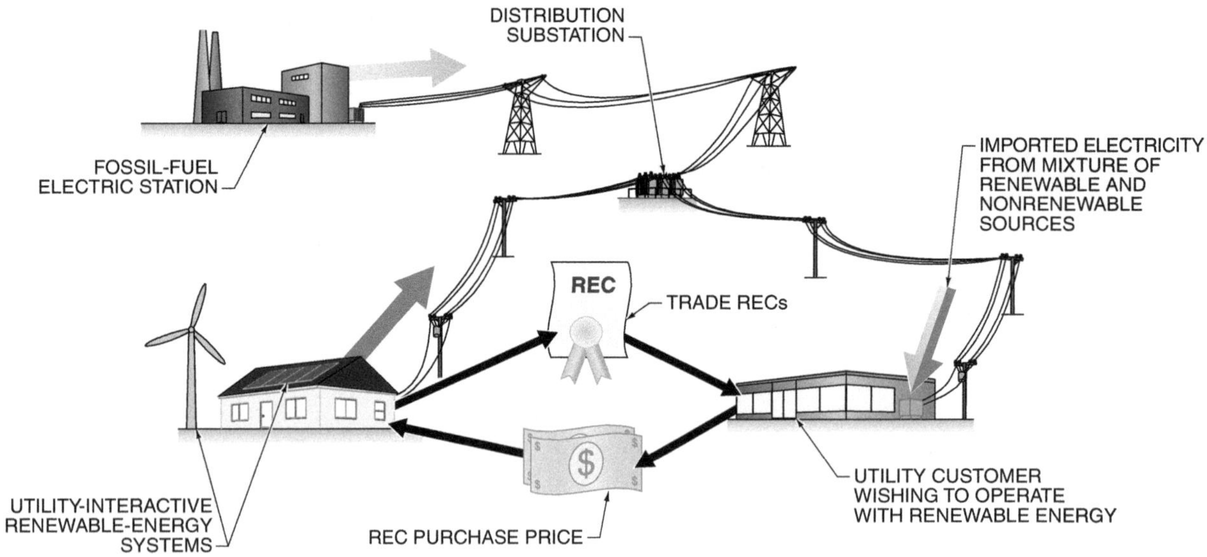

1. How does an REC work?

2. What must be carefully controlled in order to make this system work?

3. What are the advantages of this system to the REC seller and buyer?

Activity: Energy Audits

Answer the following questions about energy audits.

1. What are the differences between the different types of energy audits?

2. Which type of energy audit is most cost-effective?

3. What factors are typically investigated during an energy audit?

4. Why does an investment-grade energy audit require significant investment in modeling energy use and estimating savings?

Activity: Motor Sizing

The graph shows motor efficiency for different load factors.

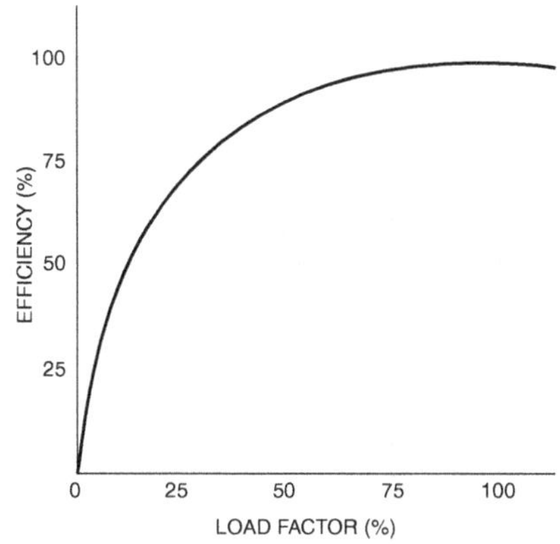

1. What are the effects of an overloaded motor and an underloaded motor?

2. At what load factors do most motors operate most efficiently?

3. How is motor load factor determined?

4. What should be done about underloaded motors?

Activity: Refrigeration System Setpoints

The maintenance personnel at a facility is considering changing the setpoint for a refrigeration system.

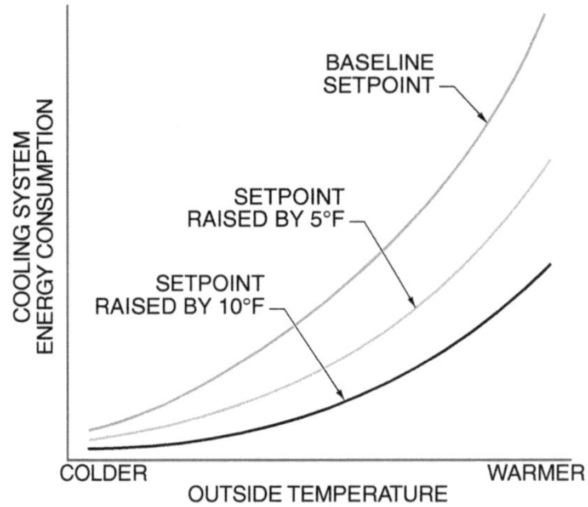

1. Describe how changing the setpoint affects energy consumption.

2. Why does the energy-use profile depend on the outside temperature?

3. What must be considered before changing setpoints?

Activity: Load Balancing

The graph shows the average power demand from a utility for 24 hours.

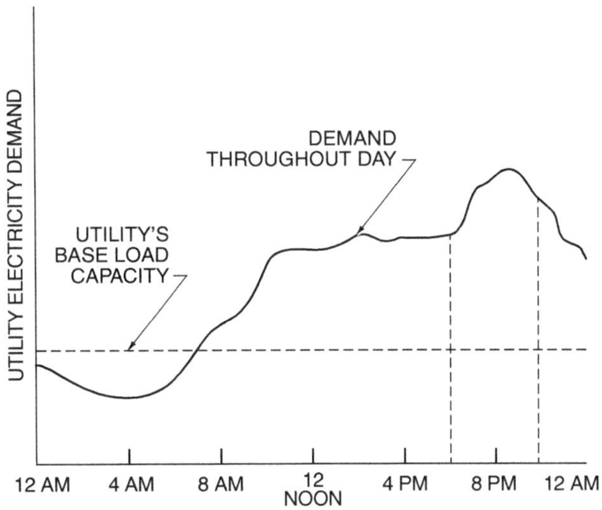

1. What times of day will likely have the highest energy price, and what times of day may have the lowest price?

2. If the utility uses time-of-use pricing, how can a facility reduce its energy costs?

3. How does load shedding reduce energy use during peak times?

4. How does load shifting reduce energy costs?

5. What types of equipment controls would be needed for load balancing strategies?

Activity: Pneumatic System Leaks

The table shows average leakage rates for different leak sizes and system pressures.

Leakage Rates*						
Pressure†	Hole Size‡					
	1/64	1/32	1/16	1/8	1/4	3/8
70	0.3	1.2	4.8	19.2	76.7	173
80	0.33	1.3	5.4	21.4	85.7	193
90	0.37	1.5	5.9	23.8	94.8	213
100	0.41	1.6	6.5	26.0	104	234
125	0.49	2.0	7.9	31.6	126	284

* in cfm
† in psig
‡ in in.

1. How does system pressure affect leakage rates?

2. What are some ways to find pneumatic system leaks?

3. If a system has several leaks of different sizes, which should be fixed first?

4. What affect would significant leakage have on the compressor?